HIGH TEMPERATURE SUPERCONDUCTIVITY

HIGH TEMPERATURE SUPERCONDUCTIVITY

The First Two Years

Proceedings of a Conference held
11–13 April 1988
Tuscaloosa, Alabama

edited by

Robert M. Metzger

Department of Chemistry
University of Alabama, Tuscaloosa

GORDON AND BREACH SCIENCE PUBLISHERS
New York London Paris Montreux Tokyo Melbourne

© 1989 by OPA (Amsterdam) B.V. All rights reserved. Published under license by Gordon and Breach Science Publishers S.A.

Gordon and Breach Science Publishers

Post Office Box 786
Cooper Station
New York, New York 10276
United States of America

Post Office Box 197
London WC2E 9PX
England

58, rue Lhomond
75005 Paris
France

Post Office Box 161
1820 Montreux 2
Switzerland

3-14-9, Okubo
Shinjuku-ku, Tokyo
Japan

Private Bag 8
Camberwell, Victoria 3124
Australia

Library of Congress Cataloging-in-Publication Data

High temperature superconductivity : the first two years / edited by
Robert M. Metzger.
        p.   cm.
    ''Proceedings of the International Conference on the First Two
Years of High-Temperature Superconductivity (Bryant Convention
Center, University of Alabama, Tuscaloosa, Alabama, 11–13 April
1988)''—Pref.
    Includes index.
    ISBN 2-88124-299-5
    1. High temperature superconductivity—Congresses. 2. High
temperature superconductors—Congresses. I. Metzger. R. M. (Robert
M.). 1940–   . II. International Conference on the First Two Years
of High-Temperature Superconductivity (1988 : Bryant Convention
Center, University of Alabama)
QC611.98.H54H544 1988   88-28468
537.6′23—dc19   CIP

CONTENTS

THE 90 K FRONTIER - INVITED PAPERS

THE 40 K FRONTIER - CONTRIBUTED PAPER

THE 90 K FRONTIER - CONTRIBUTED PAPERS

THE 120 K FRONTIER - CONTRIBUTED PAPERS

PROCESSING - INVITED PAPERS

PROCESSING - CONTRIBUTED PAPERS

THEORY - INVITED PAPERS

THEORY - CONTRIBUTED PAPERS

PREFACE

This volume contains the proceedings of the International Conference on the First Two Years of High-Temperature Superconductivity (Bryant Convention Center, University of Alabama, Tuscaloosa, Alabama, 11-13 April 1988).

The idea for the conference originated in July 1987 with Mr. Charles Forman of the College of Continuing Studies, University of Alabama, and with Dr. Donald A. Stanley of the U. S. Bureau of Mines, Tuscaloosa Research Station. The paper that pushed the highest-known superconducting critical temperature, T_c, well beyond its 1973 record of 23 K, was submitted to Zeitschift für Physik in April 1986 by Johann Georg Bednorz and Karl Alex Müller of IBM Zürich Research Laboratory; this result was confirmed by other groups in November 1986. The rapid advance in critical temperatures beyond 90 K occurred almost simultaneously in Huntsville, Alabama, in Tokyo, Japan, and in Beijing, China in January 1987.

Thus, it seemed fitting to celebrate the first Zürich result two years after the submission of the now world-famous article by Bednorz and Müller, and to attract a small but well-chosen group of speakers and participants to Alabama, the state where Wu had broken through the "77 K barrier" of superconducting critical temperatures. Prof. Müller was one of the early invited guests of the Tuscaloosa Conference, but, later, under the pressures generated by the award of the well-deserved Nobel Prize to him and to Dr. Bednorz, had to withdraw. In the mean time, the field exploded with research efforts world-wide, publications, and, alas...conferences.

Recently, there have been conferences on high-temperature superconductivity every month, sometimes every two weeks, somewhere in the world. Some of the speakers are continuously on the run; one invitee declined my invitation with "if I go to one more meeting, my wife will kill me." In the hurly-burly of such meetings, there seemed to be a need for a small, intimate conference, where one could sit and think together.

The Tuscaloosa meeting was sponsored by the University of Alabama and the U. S. Bureau of Mines. It was organized with the help and counsel of Dr. Stanley, and Profs. Anny Morrobel-Sosa, Chester Alexander, Jr., and Doru Stefanescu. The staff of Mr. Forman, especially Mrs. Nova Hodo, was instrumental with the mechanics of the conference. Mrs. Christina Metzger entertained the spouses of the conferees. The conference received financial support from the U. S. Bureau of Mines, the U. S. Office of Naval Research, the Alabama Dept. of Economic Development and Community Affairs, and Anderson Physics Laboratories, Inc. Some of the clerical work was done by Mrs. Evelyn Jackson. I thank all these people for their invaluable contributions; I thank also the referees for their speedy yet painstaking review of all the manuscripts, and Apple Inc. for inventing the MacIntosh computer.

The Tuscaloosa meeting consisted of 19 invited speakers, 21 contributed posters, plus three contributed papers sent later *in absentia* (Momin, Grigoryan, Kulik), and 118 participants. The speakers included not only Wu, but also Raveau (on whose solid-state chemistry much of the work on ceramic superconductors is founded) Hermann (who broke through the 120 K barrier), Parkin (whose work holds the present record T_c (zero resistance) = 125 K), Emery (whose theoretical views are getting wide acceptance). We were also honored by the distinguished presence, and lively participation of Professor Linus Pauling, 87 years young. No earth-shaking results were announced in Tuscaloosa (this cannot be mandated in advance) but we did hear a full and detailed account of the large and controversial isotope effect work by Ott and Smith *et al.* at Los Alamos.

We heard in sad detail from Sorrell about the dearth of new federal research support for high-temperature superconductivity in the U.S., despite the 1987 political hoopla. Mr. Ron Dagani took many photographs, and obtained useful data for his thoughtful review of the field of high-temperature superconductivity, which appeared in the 16 May 1988 issue of *Chemical and Engineering News.*

Crucial advances in this field were made in small laboratories with limited budgets. As Emery has said elsewhere, this is "the triumph of small physics;" Paul Chu had called it the "triumph of third-world physics". The great gains in superconducting critical temperatures were made in a small IBM laboratory in Zürich, Switzerland (which produced two Nobel prizes in two successive years!), in an even smaller laboratory in Huntsville, Alabama, in a large but not very well-known laboratory in Tsukuba, Japan, and in a small laboratory in Fayetteville, Arkansas. The breakthroughs gave a lot of work for the larger laboratories, and hope for future successes for the smaller ones.

There is a sobering minority report in these Proceedings by Norton. The new superconducting ceramics are no panacea. But nobody should lose heart. Even if the breakthroughs in critical temperature, densification, and critical current are obtained, the great promised applications of high-temperature superconductivity may still require years of painstaking applied technological effort.   Thus, the work on high-temperature superconductors will require some money, large amounts of patience and dedication, and good ideas.   In the early 1930's David Harker asked his advisor, "Dr. Pauling, how does one get good ideas?"   Pauling answered, "Well, I suppose one gets lots of ideas, and throws away the bad ones."

                              Robert M. Metzger
                              Tuscaloosa

1. R. G. Goodrich
2. T. Siegrist
3. R. M. Metzger
4. L. N. Mulay
5. M. H. Cohen
6. S. S. P. Parkin
7. B. Yarar
8. A. M. Hermann
9. L. Pauling
10. A. Morrobel-Sosa
11. C. A. Sorrell
12. B. Raveau
13. M. K. Wu
14. R. E. Clapp
15. X.-P. Shen
16. W. Mi
17. M. A. Takassi
18. B. I. Lee
19. D. A. Robinson
20. A. Soeta
21. C. Alexander, Jr.
22. K. Sekizawa

23. J.-M. Tarascon
24. S. Sen
25. C. Asavaroengchai
26. M. Attia
27. H. Tang
28. E. A. Boudreaux
29. A. F. Hepp
30. L. Oden
31. R. B. King
32. M. L. Norton
33.
34. L. A. Curtiss
35. J. O. Crabtree
36. T. Hasegawa
37. C.-H. Chen
38. B. M. White
39. J. James
40. F. A. Morse
41.
42. C. C. Torardi
43. J. L. Smith
44. C. M. Sung

45. I.-G. Chen
46. D. W. Capone II
47. A. Maginnis
48. J. Debsikdar
49. D. A. Stanley
50. T. Matsubara
51. T. R. Brumleve
52.
53. P. G. Wahlbeck
54.
55. W. E. Hatfield
56. R. W. McCallum
57. S. K. Mendiratta
58. N.-P. Ong
59. D. G. Agresti
60. E. Gattef
61. D. W. Cooke
62. I. Zitkovski
63. D. Lianos
64.
65. J. P. Franck
66. A. J. Epstein

PROGRAM

SUNDAY 10 APRIL 1988

12:00 noon to   REGISTRATION, Bryant Conference Center
8:30 p.m.

==

MONDAY 11 APRIL 1988

9:00 a.m. to   **SESSION A**, Bryant Conference Center
12:20 p.m.   **Chair:** Dr. Charles A. **Sorrell** (U. S. Bureau
                 of Mines)

9:00 a.m.   FORMAL OPENING
                 Prof. Joab **Thomas**
                 President of The University of Alabama

9:15 a.m.   INTRODUCTORY REMARKS
                 Prof. Robert M. **Metzger**

9:20 a.m.   A1. Prof. Ching-Wu "Paul" **Chu** (Univ. of
                 Houston)
                 High-Temperature Superconductivity - Past,
                 Present, and Future! *(withdrawn)*

10:00 a.m.   A2. Prof. Bernard **Raveau** (Univ. de Caen,
                 France)
                 Role of Oxygen and Structural Defects in
                 the Properties of High-T_c Superconductors

11:00 a.m.   A3. Dr. Tetsuya **Hasegawa** (Univ. of Tokyo,
                 Japan)
                 Oxygen Deficiency and its Effect on
                 Superconductivity and Electrical Properties

11:40 a.m.   A4. Prof. Maw-Kuen **Wu** (Univ. of Alabama in
                 Huntsville)
                 Recent Developments in High-T_c Oxide
                 Superconductors at the University of Alabama
                 in Huntsville

7:30 p.m. to   **SESSION B**, Bryant Conference Center
9:40 p.m.   **Chair**: Prof. Maw-Kuen **Wu** (Univ. of Alabama
                 in Huntsville)

7:30 p.m.   B1. Dr. James L. **Smith** (Los Alamos Natl. Lab.)
                 A Large Isotope Shift in $YBa_2Cu_3O_{7-x}$

8:10 p.m.   B2. Prof. Nai-Phuan **Ong** (Princeton Univ.)
                 Transport and Tunneling Experiments on Single Crystals of $YBa_2Cu_3O_7$

8:50 p.m.   B3. Prof. Arthur J. **Epstein** (Ohio State Univ.)
                 Anomalous Magnetic, Photo-Induced, and Structural Phenomena in High-T_c Materials

9:30 p.m.   B4. Prof. Laxman N. **Mulay** (Pennsylvania State Univ.)
                 Inexpensive Classical Techniques for Magnetic Measurements on High-T_c Superconductors

----

9:40 p.m. to   **POSTER SESSION**, Bryant Conference
11:00 p.m.   Center

PROCESSING POSTERS:

P1.   Prof. David E. **Farrell** (Case Western Reserve Univ.)
       Doping Directed at the Oxygen Sites in $YBa_2Cu_3O_{7-\delta}$: the Effect of Sulfur, Fluorine, and Chlorine

P2.   Dr. Aloysius F. **Hepp** (NASA Lewis Res. Ctr.)
       Synthesis and Processing Study of Ceramic Superconductors: Inert Atmospheres, Alternative Precursors,and Limits on J_c

P3.   Prof. Michael L. **Norton** (Univ. of Georgia)
       The Myth of the Possibility of High Critical Current in Micro-crystalline Oxide Wires: The Failure of Three-Dimensional Processing Techniques in Low Dimensions

P4.   Prof. Chang-Mo **Sung** (Lehigh Univ.)
       Superconducting Oxide Processing in the Bi-Ca-Sr-Cu-O System: Characterization and Comparison by Analytical Electron Microscopy

P5.   Ms. Darla **Schrodt** (Naval Res. Lab.)
       Processing and Characterization in the Bi-Sr-Ca-Cu-O
       System

P6.   Prof. Burtrand I. **Lee** (Clemson Univ.)
       Processing of Flexible High-T_c Superconducting Wires

P7.   Dr. N. K. **Jaggi** (Northeastern Univ.)
       Deposition of Thin Films of Bi-Sr-Ca-Cu-Oxide
       Superconductors by Laser Ablation *(withdrawn)*

P8.   Dr. Emil M. **Gattef** (Higher Inst. of Chem. Technology,
       Sofia, Bulgaria)
       Anomalous Superconductivity in the System
       Y-Ba-Cu-Ag-O

P9.   Prof. Anny **Morrobel-Sosa** (Univ. of Alabama)
       Studies on Ozone Processing of High-T_c
       Superconductors

P10.   Dr. In-Gann **Chen** (Univ. of Alabama)
       Experiments on the Growth of Y-Ba-Cu Oxide Single
       Crystals

CHARACTERIZATION POSTERS:

P11.   Dr. A. C. **Momin** (Bhabha Atomic Res. Ctr., Bombay,
       India)
       Bulk Thermal Expansion Studies of the Superconducting
       $YBa_2Cu_3O_7$ *(in absentia)*

P12.   Prof. David G. **Agresti** (Univ. of Alabama at Birmingham)
       Mössbauer Studies of High-T_c Superconductors
       (withdrawn)

P13.   Prof. P. **Boolchand** (Univ. of Cincinnati)
       Mössbauer Debye-Waller Factors in High-T_c
       Superconductors
       (presented by I. Zitkowski)

P14.   Dr. D. W. **Cooke** (Los Alamos Natl. Lab.)
       Positive Muon Spin Depolarization in $MBa_2Cu_3O_x$ (M
       =Gd, Eu)

P15. Prof. William E. **Hatfield** (Univ. of North Carolina)
Synthesis, Characterization, and Magnetic Properties of
Bi-Sr-Ca-Cu-O_x Ceramics, a New Class of High-T_c
Superconductors

P16. Prof. Kazuko **Sekizawa** (Nihon Univ., Tokyo, Japan)
Superconductivity and Magnetism of La-Cu-O
Compounds with the K_2NiF_4 Structure

P17. Prof. Dilip K. **De** (Bowman Gray School of Medicine)
Determination of Site Symmetry of Cu(II) Ion, Critical
Fields, and London Penetration Depth from ESR and IR
Studies of the High-T_c Superconductor $YBa_2Cu_3O_{7-x}$
(withdrawn)

P18. Prof. R. G. **Goodrich** (Louisiana State Univ.)
Radio-Frequency Surface Impedance Measurements on
$YBa_2Cu_3O_{7-x}$

THEORY POSTERS:

P19. Prof. R. Bruce **King** (Univ. of Georgia)
Chemical Bonding Topology of Superconductors

P20. Prof. Edward A. **Boudreaux** (Univ. of New Orleans)
A Molecular Orbital Study of Bonding in $YBa_2Cu_3O_7$

P21. Dr. Larry A. **Curtiss** (Argonne Natl. Lab.)
Theoretical Studies of Copper Oxide Clusters:
Investigation ofCu(III) in $YBa_2Cu_3O_{7-x}$

P22. Dr. Roger E. **Clapp** (MITRE Corp.)
Charge Circulation in High-T_c Superconductors

P23. Prof. Sanjoy K. **Sarker** (Univ. of Alabama)
New Functional Integral Theory for the Strongly
Correlated Hubbard Model

P24. Prof. Jürgen P. **Franck** (Univ. of Alberta).

P25. Mr. Toshiya **Matsubara** (Asahi Glass Co.)

===

TUESDAY 12 APRIL 1988

8:30 a.m. to **SESSION C**, Bryant Conference Center
12:30 p.m.   Chairs: Prof. Bernard **Raveau** (Univ. deCaen) and
             Prof. R. Bruce **King** (Univ. of Georgia)

8:30 a.m.   C1. Dr. Jean-Marie **Tarascon** (Bellcore)
             On the Synthesis, Structural, and Physical
             Properties of the High-T_c $YBa_2Cu_{3-x}M_xO_{7-y}$

9:10 a.m.   C2. Prof. Allen M. **Hermann** (Univ. of Arkansas)
             120K Tl-Ca-Ba-Cu-O Bulk Superconductors

9:30 a.m.   C3. Dr. Stuart S. P. **Parkin** (IBM Almaden Res. Ctr.)
             Superconductivity at 125K in $Tl_2Ca_2Ba_2Cu_3O_x$

10:30 a.m.   C4. Dr. Theo **Siegrist** (AT&T Bell Labs.)
             Precision Measurements on High-T_c
             Superconductors

11:10 a.m.   C5. Dr. Charles C. **Torardi** (Du Pont Central Res.)
             Synthesis and Structure-Property Relationships in
             the Bi-Sr-Ca-Cu-O and Tl-Ba-Ca-Cu-O
             Superconductors

11:50 a.m.   C6. Dr. Jagedish **Debsikdar** (Idaho Natl. Eng. Lab.)
             Sol-Gel Routes to Superconducting Ceramics

7:30 p.m. to **SESSION D**, Bryant Conference Center
9:30 p.m.   **Chair**: Prof. Doru M. **Stefanescu** (Univ. of
             Alabama)

7:30 p.m.   D1. Prof. R. William **McCallum** (Iowa State Univ.)
             Microstructural Considerations in Polycrystalline
             $YBa_2Cu_3O_7$

8:10 p.m.   D2. Prof. Baki **Yarar** (Colorado School of   Mines)
             Application of Principles of Process Metallurgy to the
             Production of Copper-Sheated Y-Ba-Cu-O Wires

8:50 p.m.   D3. Session on Funding : Dr. Charles **Sorrell**
             (Bureau of Mines)

9:30 p.m. to **POSTER SESSION**, Bryant Conference
11:00 p.m.   Center Same posters as Monday night

WEDNESDAY 13 APRIL 1988

8:30 a.m. to 12:10 p.m.	**SESSION E**, Bryant Conference Center **Chair**: Prof. William E. **Hatfield** (Univ. of North Carolina)
8:30 a.m.	E1. Dr. Donald W. **Capone** II (Argonne Natl. Lab.) Low-Temperature Densification of High-T_c Superconductors
9:10 a.m.	E2. Dr. Stuart **Wolf** (Naval Res. Lab.) What's Hot in Superconductivity at NRL
9:50 a.m.	E3. Prof. Linus **Pauling** (Pauling Inst.) The Role of the Metallic Orbital and of Crest and Trough Superconduction in High-Temperature Superconductors
10:50 a.m.	E4. Dr. Victor J. **Emery** (Brookhaven Natl. Lab.) Magnetism and Superconductivity in High-T_c Superconductors
11:30 a.m.	E5. Dr. Morrel H. **Cohen** (Exxon Res.) Phenomenology of High-Temperature Superconductors

==

PARTICIPANTS

(Invited speakers in **boldface**; principal poster presenters underlined)

1. Prof. David G. Agresti, Dept. of Physics, Univ. of Alabama at Birmingham, Birmingham, AL 35294, USA. (205)-934-4736.
2. Mr. Walid Al-Akhdar, Dept. of Chemistry, Univ. of Alabama, Tuscaloosa, AL 35487, USA. (205)-348-5954.
3. Prof. Chester Alexander, Jr., Dept. of Physics, Univ. of Alabama, Tuscaloosa, AL 35487, USA. Tel (205)-348-3780.
4. Mr. Chinnarong Asavaroengchai, Dept. of Chemistry, Univ. of Alabama, Tuscaloosa, AL 35487, USA. (205)-348-5954.
5. Mr. Mohamed Attia, Dept. of Physics, Univ. of Alabama, Tuscaloosa, AL 35487, USA. (205)-348-5050.
6. Mr. Mazla A. Bakar, Dept. of Physics, Univ. of Alabama, Tuscaloosa, AL 35487, USA. (205)-348-5050.
7. Prof. John T. Berry, Dept. of Metallurgical Eng., Univ. of Alabama, Tuscaloosa, AL 35487. (205)-348-1747.
8. Prof. Edward A. Boudreaux, Dept. of Chemistry, Univ. of New Orleans, New Orleans, LA 70148, USA. (504)-286-6311.
9. Mr. Eric G. Bradford, Dept. of Chemistry, Univ. of Alabama, Tuscaloosa, AL 35487, USA. (205)-348-5954.
10. Dr. Timothy R. Brumleve, Anderson Physics Laboratories, Inc., 406 N. Busey Avenue, Urbana, IL 61801, USA. (217)-356-1347.
11. Ms. Mitzie Burleson, Dept. of Chemistry, Univ. of Alabama, Tuscaloosa, AL 35487, USA. (205)-348-5954.
12. Dr. Donald W. **Capone** II, Materials Science Division, 23-A113, Argonne National Laboratory, Argonne, IL 60439, USA. (312)-972-5526.
13. Mr. James C. Cates, Dept. of Physics, Univ. of Alabama, Tuscaloosa, AL 35487, USA. (205)-348-5050.
14. Dr. Bryan C. Chakoumakos, Solid State Division, Oak Ridge National Laboratory, Bldg. 2000, Box X, Oak Ridge, TN 37830-0656, USA. (615)-574-5495.
15. Dr. In-Gann Chen, Dept. of Metallurgical Engineering, Univ. of Alabama, Tuscaloosa, AL 35487, USA. (205)-345-1748.
16. Mr. Andre Chiu, Dept. of Chemistry, Univ. of Alabama, Tuscaloosa, AL 35487, USA. (205)-348-5954.
17. Mr. Timothy Clancy, U. S. Bureau of Mines, Tuscaloosa, AL 35486, USA. (205)-759-9400.
18. Dr. Roger E. Clapp, The MITRE Corporation, Mail Stop N225, Burlington Road, Bedford, MA 01730, USA. (617)-271-5622.
19. Dr. Morrel H. **Cohen**, Exxon Research and Engineering Co., Clinton Township, Route 22 East, Annandale, NJ 08801, USA. (201)-730-2439.

20. Dr. D. Wayne Cooke, Los Alamos National Laboratory, MP14, MS H844, Los Alamos, NM 87545, USA. (505)-667-4274.
21. Mr. Lalith M. Cooray, Dept. of Chemistry, Univ. of Alabama, Tuscaloosa, AL 35487, USA. (205)-348-5954.
22. Mr. Jerome O. Crabtree, 1101 13th Street, Tuscaloosa, AL 35401.
23. Mr. Peter J. Cragg, Dept. of Chemistry, Univ. of Alabama, Tuscaloosa, AL 35487, USA. (205)-348-5954.
24. Dr. Larry A. Curtiss, Chemical Technology Division, Argonne National Laboratory, Argonne, IL 60439, USA. (312)-972-7380.
25. Mr. Ron Dagani, Chemical and Engineering News, 1155 16th Street NW, Washington, DC 20036, USA. (202)-872-4488.
26. Dr. Jagadish C. **Debsikdar**, Materials Technology, Idaho National Engineering Laboratory, EG & G Idaho, Inc., Mail Stop 2210, Idaho Falls, ID 83415, USA. (208)-526-8016.
27. Prof. B. K. Dhindaw, Dept. of Metallurgical Engineering, Univ. of Alabama, Tuscaloosa, AL 35487, USA. (205)-348-1740.
28. Mr. Ruisong Ding, Dept. of Chemistry, Univ. of Alabama, Tuscaloosa, AL 35487, USA. (205)-348-5954.
29. Dr. Victor J. **Emery**, Dept. of Physics, Brookhaven National Laboratory, Upton, NY 11973, USA. (516)-282-3765.
30. Prof. Arthur J. **Epstein**, Dept. of Physics, Ohio State Univ., Columbus, OH 43210-1106, USA. (614)-292-1133.
31. Prof. David E. Farrell, Dept. of Physics, Case Western Reserve Univ., Cleveland, OH 44106, USA. (216)-368-2615.
32. Mr. Charles Forman, Assoc. Director, Environmental and Industrial Programs, College of Continuing Studies, Univ. of Alabama, Tuscaloosa, AL 35487, USA. (205)-348-9718.
33. Prof. Jürgen P. Franck, Dept. of Physics, Univ. of Alberta, Edmonton, Alberta T6G 2J1, Canada. (403)-432-4529.
34. Dr. Emil Gattef, Central Research Laboratory, Higher Institute of Chemical Technology, Blv. K. Ohridsky 8, Sofia 1156, Bulgaria. 359-(2)-625-385.
35. Mr. Barry Glover, Dept. of Chemistry, Univ. of Alabama, Tuscaloosa, AL 35487, USA. (205)-348-5954.
36. Dr. R. G. Goodrich, Dept. of Physics and Astronomy, Louisiana State Univ.,Baton Rouge, LA 70803, USA. (504)-388-5847.
37. Mr. Yu Guan, Dept. of Physics, Univ. of Alabama, Tuscaloosa, AL 35487, USA. (205)-348-5050.
38. Mr. Kyoung Han, Bureau of Mines, Tuscaloosa Research Center,Tuscaloosa, AL 35487, USA. (205)-759-9400.
39. Professor James W. Harrell, Jr., Dept. of Physics, Univ. of Alabama, Tuscaloosa, AL 35487, USA. (205)-348-3785.

40. Dr. Tetsuya **Hasegawa**, Dept. of Industrial Chemistry, Univ. of Tokyo, Hongo, Bunkyo-ku, Tokyo 113, Japan. 81-(3)-812-2111 Ext. 7204.
41. Professor William E. Hatfield, Dept. of Chemistry, Univ. of North Carolina, Chapel Hill, NC 27514, USA. (919)-966-2297.
42. Dr. Aloysius F. Hepp, NASA Lewis Research Center, MS 302-1, 21000 Brookpark Road,Cleveland, OH 44135, USA. (216)-433-3835.
43. Professor Allen M. **Hermann**, Dept. of Physics, Univ. of Arkansas, Fayetteville, AR 72701, USA.   (501)-575-2506.
44. Mr. Ichiro Ishida, Microelectronics Research Labs., NEC Corporation, 1-1 Miyazaki 4-chome, Miyamae-ku, Kawasaki, Kanagawa 213, Japan. 81-(44)-855-1111
45. Mr. Sahlan Jamal, Dept. of Physics, Univ. of Alabama, Tuscaloosa, AL 35487, USA. (205)-348-5050.
46. Mr. John C. James, Dept. of Physics and Astronomy, Univ. of Alabama,Tuscaloosa, AL 35487, USA. (205)-348-5050.
47. Mr. Thomas K. Jones, Jr., Dept. of Chemistry, Univ. of Alabama, Tuscaloosa, AL 35487, USA. (205)-348-5954.
48. Mr. Makarand V. Joshi, Dept. of Chemistry, Univ. of Alabama, Tuscaloosa, AL 35487, USA. (205)-348-5954.
49. Mr. Wayne Keen, Dept. of Physics, Univ. of Alabama, Tuscaloosa, AL 35487, USA. (205)-348-5050.
50. Professor Maureen Kendrick, Dept. of Chemistry, Univ. of Alabama, Tuscaloosa AL 35487, USA. (205)-348-9203.
51. Professor Robert Bruce King, Dept. of Chemistry, Univ. of Georgia, Athens, GA 30602, USA. (404)-542-1901.
52. Professor Lowell D. Kispert, Dept. of Chemistry, Univ. of Alabama,Tuscaloosa, AL 35487, USA. (205)-348-7134.
53. Professor Burtrand I. Lee, Ceramic Eng. Dept., Clemson Univ., Clemson, SC 29634-0907, USA. (803)-656-2834.
54. Mr. Dimitrios Lianos, J. W. Schaefer Co., 1500 Perimeter Parkway, Suite 275, Huntsville, AL 35806, USA. (205)-721-9572.
55. Mr. Zheng-Heng Lin, Dept. of Physics, Univ. of Alabama, Tuscaloosa, AL 35487, USA. (205)-348-5050.
56. Mr. Qian-Cheng Ma, Dept. of Physics, Univ. of Alabama, Tuscaloosa, AL 35487, USA. (205)-348-5050.
57. Mr. Krishnan Maddapat, Dept. of Chemistry, Univ. of Alabama, Tuscaloosa, AL 35487, USA. (205)-348-5954.
58. Mr. M. Abbot Maginnis, U. S. Bureau of Mines, Tuscaloosa Research Center,Tuscaloosa, AL 35487-9777, USA. (205)-759-9400.
59. Mr. Ekramul Majid, Dept. of Physics, Univ. of Alabama, Tuscaloosa, AL 35487, USA. (205)-348-5050.

60. Mr. Hiroshi Makino, Dept. of Physics, Univ. of Alabama
    Tuscaloosa, AL 35487, USA.   (205)-348-5050.
61. Mr. Toshiya Matsubara, Research and Development Division,
    Asahi Glass Co. Ltd., Hazawa-cho, Kanagawa-ku, Yokohama
    221, Japan.
62. Mr. Michael May, Dept. of Chemistry, Univ. of Alabama,
    Tuscaloosa, AL 35487, USA. (205)-348-5954.
63. Prof. R. William **McCallum**, Dept. of Metallurgy, 106 Wilhelm,
    Ames Laboratory USDOE, Iowa State Univ., Ames, IA 50011,
    USA. (515)-294-4736.
64. Dr. Sudhir K. Mendiratta, Olin Corp., P. O. Box 248,
    Charleston, TN 37310, USA.
65. Prof. Robert M. Metzger, Dept. of Chemistry, Univ. of
    Alabama, Tuscaloosa, AL 35487, USA.   (205)-348-5952.
66. Ms. Wei Mi, Dept. of Physics, Univ. of Alabama, Tuscaloosa
    AL, 35487, USA. (205)-348-5050.
67. Prof. Ichiro Miyagawa, Dept. of Physics, Univ. of Alabama,
    Tuscaloosa, AL 35487. (205)-348-3787.
68. Professor Anny Morrobel-Sosa, Dept. of Chemistry, Univ. of
    Alabama, Tuscaloosa, AL 35487, USA. (205)-348-9116.
69. Dr. Fred A. Morse, Los Alamos National Laboratory, Mail
    Stop A114, P. O. Box 1663, Los Alamos, NM 87545, USA.
    (505)-667-1600.
70. Professor Laxman N. Mulay, Dept. of Materials Science and
    Eng., Pennsylvania State Univ., Univ. Park, PA 16802, USA.
    (814)- 865-1621.
71. Ms. Renee Niziurski, Dept. of Chemistry, Univ. of Alabama,
    Tuscaloosa, AL 35487, USA. (205)-348-5954.
72. Prof. Michael L. Norton, Dept. of Chemistry, Univ. of Georgia,
    Athens, GA 30602, USA.   (404)-542-2023.
73. Dr. Laurance L. Oden, U. S. Bureau of Mines, Albany
    Research Center, Albany, OR 97321, USA. (503)-967-5862.
74. Ms. Sandra Oldham, Dept. of Chemistry, Univ. of Alabama,
    Tuscaloosa, AL 35487, USA. (205)-348-5954.
75. Professor Nai-Phuan **Ong**, Dept. of Physics, P.O. Box 708,
    Princeton Univ., Princeton, NJ 08544, USA. (609)-452-4347.
76. Mr. Bill Orr, Dept. of Chemistry, Univ. of Alabama,
    Tuscaloosa, AL 35487, USA. (205)-348-5954.
77. Mr. Shih-Jyun Pan, Dept. of Physics, Univ. of Alabama,
    Tuscaloosa, AL 35487, USA. (205)-348-5050.
78. Dr. Stuart S. P. **Parkin**, IBM Almaden Research Center,
    K32/803(D), 650 Harry Road,San Jose, CA 95120-6099,
    USA. (408)-927-2390.
79. Prof. Linus C. **Pauling**, Linus Pauling Institute, 440 Page Mill
    Road, Palo Alto, CA 95306, USA. (415)-327-4064.
80. Mr. Martin J. Plishka, Dept. of Chemistry, Univ. of
    Alabama,Tuscaloosa, AL 35487, USA. (205)-348-5954.

81. Dr. Lakkaraju S. Prasad, Dept. of Chemistry, Univ. of Alabama, Tuscaloosa, AL 35487, USA. (205)-348-5954.
82. Prof. Bernard **Raveau**, Laboratoire de Cristallographie, Chimie et Physique des Solides, UA251 ISMRa (CNRS), Ecole Nationale Superieure d'Ingenieurs, Boulevard du Marechal Juin, F-14032 Caen Cedex, France. 33-(31)-45-2616.
83. Mr. David A. Robinson, Dept. of Chemistry, Univ. of Alabama, Tuscaloosa, AL 35487, USA. (205)-348-5954.
84. Mr. Kerry D. Robinson, Dept. of Chemistry, Univ. of Alabama, Tuscaloosa, AL 35487, USA. (205)-348-5954.
85. Prof. Sanjoy K. Sarker, Dept. of Physics, Univ. of Alabama, Tuscaloosa, AL 35487, USA. (205)-348-3772.
86. Ms. Darla Schrodt, Ceramic Eng., Code 6371, Naval Research Laboratory,Washington, DC 20375-5000, USA. (202)-767-2007.
87. Prof. Kazuko Sekizawa, Dept. of Physics, College of Science and Technology, Nihon Univ., 1-8 Kanda-Surugadai, Chiyoda-ku, Tokyo 101, Japan. 81- (3)-293-3251.
88. Ms. Suzanne Sellers, Dept. of Chemistry, Univ. of Alabama, Tuscaloosa, AL 35487, USA. (205)-348-5954.
89. Mr. Subhayu Sen, Dept. of Metallurgical Eng., Univ. of Alabama, Tuscaloosa, AL 35487, USA. 9205)-348-1740.
90. Ms. Xiao-Ping Shao, Dept. of Physics, Univ. of Alabama, Tuscaloosa, AL 35487, USA. (205)-348-5050.
91. Dr. Theo **Siegrist**, AT&T Bell Laboratories, 600 Mountain Avenue, Murray Hill, NJ 07974, USA. (201)-582-5253.
92. Mr. Richard Siemens, U. S. Bureau of Mines,Tuscaloosa Research Center,Tuscaloosa, AL 35487-9777, USA. (205)-759-9400.
93. Dr. James L. **Smith**, Physics Division, Los Alamos National Laboratory,Los Alamos, NM 87545, USA. (505)-667-4476.
94. Ms. Teresa Joy Smithson, Dept. of Chemistry, Univ. of Alabama, Tuscaloosa, AL 35487, USA. (205)-348-5954.
95. Ms. Atsuko Soeta, Hitachi Research Laboratory, Hitachi, Ltd., 1-1, Saiwacho 3-chome, Hitachi-shi, Ibaraki-ken 317, Japan. 81-(294)-21-111 Ext. 3126.
96. Prof. William A. Soffa, Dept. of Materials Science, University of Pittsburgh, Pittsburgh, PA 15261, USA.
97. Dr. Charles A. **Sorrell**, US Bureau of Mines - Dept. of the Interior, 2401 E Street NW, Mail Stop 4000, Washington, DC 20241, USA. (202)-634-4678.
98. Dr. Donald A. Stanley, U. S. Bureau of Mines, Tuscaloosa Research Center,Tuscaloosa, AL 35487-9777, USA. (205)-759-9455.

99. Prof. Doru M. Stefanescu, Dept. of Metallurgical Eng., Univ. of Alabama, Tuscaloosa, AL 35487, USA. (205)-348-1748.
100. Dr. Chang-Mo Sung, Dept. of Materials Science and Eng., Lehigh Univ., Bethlehem, PA 18015, USA. (215)-758-4232.
101. Dr. Koji Tada, Sumitomo Denko, 1-3 Shimaya, Konohana-ku, Osaka 554, Japan. 81-(6)-461-1031.
102. Mr. Mohammad Ali Takassi, Dept. of Chemistry, Univ. of Alabama, Tuscaloosa, AL 35487,USA. (205)-348-5954.
103. Mr. Hong Tang, Dept. of Physics, Univ. of Alabama, Tuscaloosa, AL 35487, USA. (205)-348-5050.
104. Dr. Jean-Marie **Tarascon**, Bell Communications Research, Inc., 331 Newman Springs Road, Red Bank, NJ 07701, USA. (201)-758-2930.
105. Prof. Joseph S. Thrasher, Dept. of Chemistry, Univ. of Alabama,Tuscaloosa, AL 35487, USA. (205)-348-5954.
106. Dr. Charles C. **Torardi**, Central Research, Experimental Station, E.I. du Pont de Nemours & Co., Wilmington, DE 19898, USA. (302)-695-2236.
107. Mr. Girish K. Upadhya, Dept. of Metallurgical Eng., Univ. of Alabama, Tuscaloosa, AL 35487, USA. (205)-348-1740.
108. Prof. Pieter B. Visscher, Dept. of Physics, Univ. of Alabama, Tuscaloosa, AL 35487, USA. (205)-348-3773.
109. Prof. Phillip G. Wahlbeck, Dept. of Chemistry, Wichita State Univ., Wichita, KS 67206, USA. (316)-689-3120.
110. Prof. Gary W. Warren, Dept. of Metallurgical Eng., Univ. of Alabama, Tuscaloosa, AL 35487, USA. (205)-348-1740.
111. Mr. Michael Weaver, Dept. of Chemistry, Univ. of Alabama, Tuscaloosa, AL 35487, USA. (205)-348-5954.
112. Ms. Mi Wei, Dept. of Physics, Univ. of Alabama, Tuscaloosa, AL 35487, USA. (205)-348-5050.
113. Ms. Barbara M. White, Dept. of Chemistry, Univ. of Alabama, Tuscaloosa, AL 35487, USA. (205)-348-5954.
114. Dr. Stuart A. **Wolf**, Naval Research Laboratory, Washington, DC 20375, USA. (202)-767-2600.
115. Dr. Jack Woodyard, U. S. Bureau of Mines, Tuscaloosa Research Center, Tuscaloosa, AL 35487-9777, USA. (205)-759-9400.
116. Prof. Maw-Kuen **Wu**, Dept. of Physics, Univ. of Alabama in Huntsville, Huntsville, AL 35899, USA. (205)-895-6569.
117. Prof. Baki **Yarar**, Dept. of Metallurgical Engineering, Colorado School of Mines, Golden, CO 80401, USA. (303)-273-3768.
118. Mr. Ivan Zitkovsky, Dept. of Electrical and Computer Engineering, ML20,Univ. of Cincinnati, Cincinnati, OH 45221-0030, USA. (513)-475-4461

ROLE OF OXYGEN AND STRUCTURAL DEFECTS IN PROPERTIES OF HIGH-Tc SUPERCONDUCTORS

B. RAVEAU, C. MICHEL, M. HERVIEU, F. DESLANDES
and A. MAIGNAN
Laboratoire CRISMAT, Institut des Sciences
de la Matière et du Rayonnement, Campus 2,
Bd du Maréchal Juin, 14032 Caen Cedex, France

Abstract The structures and problems of oxygen non-stoichiometry as well as defects in the new high Tc superconductors are discussed in connection with the variation of critical temperature.

Almost two years have passed since the discovery of superconductivity in mixed-valence copper oxides by Bednorz and Müller[1]. A tremendous number of papers have now been published in this field. In spite of these numerous results, the mechanisms which govern superconductivity in those oxides are so far not understood. The study of the crystal chemistry of these oxides, which started eight years ago in Caen[2], suggests that two points are of capital importance for the appearance of superconductivity in those compounds : the low dimensionality of the structure and the mixed-valence of copper.

1

THE LAYERED STRUCTURE OF THE HIGH-T_c SUPERCONDUCTORS

Six different structural types are known for their superconductivity at high temperature. The common structural feature of these different complex oxides is their relationships with the perovskite structure.

The "oldest one", is represented by the 40K-superconductors $La_{2-x}A_xCuO_{4-y}$ (A = Ca, Sr, Ba)[1-2] which belong to the K_2NiF_4 family. The structure of those oxides (Fig. 1a) corresponds to the intergrowth of superconductive single perovskite layers with insulating SrO-type layers. The oxygen deficiency of the perovskite layers is a remarkable feature which means that the Cu(III)/Cu(II) ratio may vary, leading to a modification of the superconducting properties of these oxides. La_2CuO_4, which belongs to this family, was also found to be a superconductor[2] either by changing slightly the La/Cu ratio, or by applying a high oxygen pressure.

The $YBa_2Cu_3O_{7-\delta}$ oxide called the 90K-superconductor whose superconductivity was discovered by Chu et al[3] was studied by several groups simultanously (see for instance Capponi et al[2b], Tarascon et al[3], Cava et al[4], Michel et al[2b], Schuller et al[3], Roth et al[2b]). Its structure (Fig. 1b) can be described as an ordered oxygen-deficient perovskite leading to triple layers of corner-sharing CuO_5 pyramids and CuO_4 square-planar groups containing double layers of barium cations interleaved with one layer of yttrium ions. This structural type appears very flexible, since yttrium can be replaced by lanthanides without changing either the structure or,

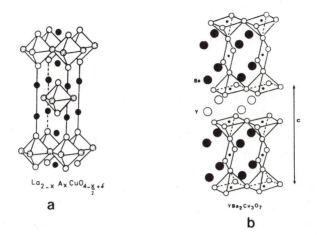

FIG. 1. a) Structure of the 40K-superconductor La$_{2-x}$Sr$_x$CuO$_{4-y}$; b) Structure of the 90K superconductor YBa$_2$Cu$_3$O$_7$.

drastically, the superconducting properties. The substitution of transition and post-transition elements for copper is also possible, but leads to a decrease of the critical temperature.

The possibility of synthesizing superconductive mixed-valence copper oxides with a layer structure, in which the lanthanides ions are replaced by bismuth, was shown for the first time in the compound "Bi$_2$Sr$_2$Cu$_2$O$_7$"[4]. From the study of the latter oxide, which exhibits a critical temperature of 22K, it was suggested that ternary copper oxides involving alkaline earth ions and bismuth could be potential materials for superconductivity, provided they are characterized by a bidimensional structure. The recent evidence for superconductivity in the range 80-105K, shown recently by Maeda et al.[5], Chu et al.[6] and Politis[7] for the composition

3

"$Bi_2 Sr_2 CaCu_2 O_8$", is in agreement with this point of view. Single crystal X-ray diffraction studies carried out by Sleight et al.[8] and by Tarascon et al.[9] allowed two structural models to be proposed for the oxide $Bi_2 Sr_2 CaCu_2 O_8$. In both models (Fig. 2) the authors agree concerning the positions of copper, strontium, calcium and oxygen surrounding the copper atoms, but are at variance about the nature of the bismuth layers ; Tarascon et al. propose a double distorted sodium chloride-type layer that will be noted here $(BiO)_2$ (Fig. 2a), whereas Sleight et al. observe a splitting of the bismuth positions (Fig. 2b) and propose $Bi_2 O_2$ layers formed of edge-sharing BiO_4 pyramids like in Aurivillius phases. High resolution electron micro-scopy and electron diffraction studies performed by Hervieu et al.[10] show the existence of satellites

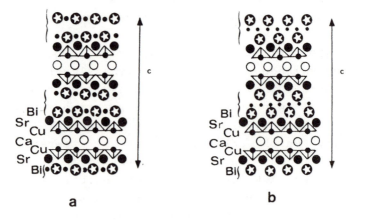

a   b

FIG. 2. Structural models of the 100K-superconduc-tor $Bi_2 Sr_2 CaCu_2 O_{8-\delta}$ along [110] according to (a) Tarascon and (b) Sleight.

4

in incommensurate positions, and suggest from careful simulations that the [BiO$_y$]$_\infty$ layers are not
[Bi$_2$O$_2$]$_\infty$ Aurivillius-type layers, but rather distorted sodium chloride-type layers. Thus the structure
of this orthorhombic phase (a . a$_p$ V2, b . a$_p$ V2, c
. 30.7 Å) can be described as built up from oxygen-
deficient perovskite layers [Sr$_2$CaCu$_2$O$_6$]$_\infty$ inter-
grown with [(BiO$_y$)$_2$]$_\infty$ layers. The [Sr$_2$CaCu$_2$O$_6$]$_\infty$
layers are themselves formed of two [CuO$_{2.5}$]$_\infty$
layers of corner-sharing pyramids similar to those
observed in YBa$_2$Cu$_3$O$_7$ and in La$_2$SrCu$_2$O$_6$[2a] ; the
cohesion between the pyramidal layers is ensured by
a plane of calcium ions, exactly in the same way
as the yttrium ions ensure the cohesion between the
[Cu$_3$O$_7$]$_\infty$ layers in YBa$_2$Cu$_3$O$_7$. The analogy with
YBa$_2$Cu$_3$O$_7$ is also observed for the order between
strontium and calcium : strontium planes alternate
with calcium planes along c.

The 22K-superconductor observed for the composition
Bi$_2$Sr$_2$Cu$_2$O$_{7+\delta}$ is assumed to have in fact the composition Bi$_2$Sr$_2$CuO$_{6+\delta}$. No serious structural study
has been carried out on this phase. However, from
its simililarity with Bi$_2$Sr$_2$CaCu$_2$O$_{8+\delta}$ and taking
into account its cell parameters (a . b . a$_p$ V2, c
. 24.4 Å) a model can be proposed (Fig. 3) corresponding to the intergrowth of single perovskite
layers [SrCuO$_3$]$_\infty$ with double distorted sodium
chloride-type [(BiO$_y$)$_2$]$_\infty$ layers.

The recent discovery of superconductivity above
100K in thallium (III) copper oxides[11] confirms the
fact that bidimensionality of the structure plays
an important role in superconductivity. Two new
superconductors were indeed identified by Hazen et

5

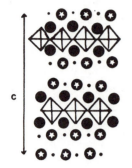

FIG. 3. Structural model proposed for the 22K-superconductor $Bi_2Sr_2CuO_6$ (stars = Bi, large black circles = Sr, small black circles = oxygen atoms in bismuth layers.

al.[12] : $Tl_2Ba_2CaCu_2O_8$ called "2212" and $Tl_2Ba_2Ca_2Cu_3O_{10}$ called "2223". Parkin et al.[13] showed that the highest T_c, of 125K was due to the "2223" phase, and attributed to the second phase "2212" a lower T_c close to 108K. However, it must be pointed out that all the thallium oxides which were studied up to now, were not obtained in the form of pure phases, due to the methods of preparation, which necessitate very short times of reaction, in order to avoid thallium oxide volatilization. In a recent study we succeeded in preparing pure compounds starting from a mixture of Tl_2O_3, CaO, BaO_2 and CuO heated in sealed tubes[14-15]. The recent study of the "2212" phase[15-16] shows that its structure (Fig. 4) is built up from infinite $[Ba_2Tl_2Cu_2O_8]_\infty$ layers whose cohesion is ensured by layers of calcium ions in the same way as calcium and yttrium are interleaved in the lamellar oxides $Bi_2Sr_2CaCu_2O_8$ and $YBa_2Cu_3O_7$. Each $[Ba_2Tl_2Cu_2O_8]_\infty$ layer is formed of layers of corner-sharing CuO_5 pyramids, and of distorted chloride-sodium type layers. Thus these $[Tl_2Ba_2Cu_2O_8]_\infty$ can be described as the intergrowths of pyramidal $[BaCuO_3]_\infty$ layers and sodium chloride type $[TlO]_\infty$ layers, according

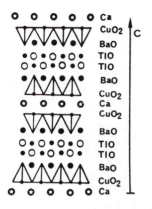

FIG.4. Idealized drawing of the structure of $Tl_2 Ba_2 CaCu_2 O_8$

to the formulation $[(BaCuO_3)_2.(TlO)_2]_\infty$. This structure is rather similar to that observed for
$Bi_2 Sr_2 CaCu_2 O_8$, the bismuth layers corresponding to
the thallium layers. However, the important difference between the two structures deals with the
very short spacing between two successive thallium
layers (2.10 Å), compared to the distance between
two consecutive bismuth layers (3.2 Å). It is worth
pointing out that no excess of oxygen can be introduced in that structure : attempts to distribute
oxygen in the calcium layers led to a significant
increase of the R factor, in agreement with the
fact that the resulting Cu-O distance would be too
short. Thus, it appears from this study that the
perfect structure $Tl_2 Ba_2 CaCu_2 O_8$ will only contain
Cu(II) and should not present any superconductivity
The electron microscopy study allows an explanation
of this phenomenon to be proposed. Numerous extended defects were indeed observed by HREM involving
thinner sodium chloride-type layers with respect to
the ideal structure $Tl_2 Ba_2 Cu_2 O_8$ (Fig. 5). Such defects can be interpreted as the intergrowth of a

7

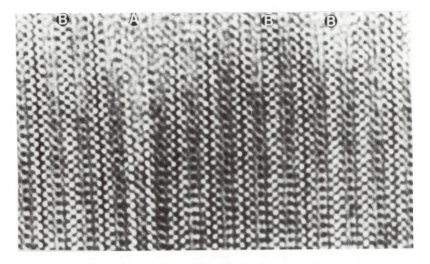

FIG. 5. HREM |010| image of the structure of a "2212" thallium oxide showing intergrowth defects.

single TlO layer with a single pyramidal layer corresponding to the sequence |Ca/CuO₂ /BaO/TlO/BaO /CuO₂ [Ca, whose local periodicity is 12.1Å. This local variation of spacing and composition corresponds to an intergrowth defect (Fig. 6) of formulation TlBa₂CaCu₂O₇ , which could involve in the area defect a high Cu(III) content (50 % of the copper sites). Such a member which exists in the form of defects, but has never been observed as a single

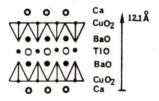

FIG. 6. Idealized drawing of the structure of the defect with composition TlBa₂CaCu₂O₇ .

phase by X-ray diffraction, could be at the origin of superconductivity. In the same way, the creation of defects on the thallium sites would lead to the same effect, corresponding also to the general formulation $Tl_{2-x}Ba_2CaCu_2O_7$. We have indeed observed an increase of the superconductive volume for the composition $TlBa_2CaCu_2O_7$ (about 30 %) in agreement with this point of view.

The structure of the "2223" phase has not yet been determined with accuracy. Nevertheless a preliminary X-ray powder study and HREM study allows a structural model to be proposed. The structure of this phase (Fig. 7a) can indeed be described as the intergrowth of a triple distorted sodium chloride type layer $[A'O]_3$ (A' = Ba, Tl) with a triple oxygen deficient perovskite layer $[ACuO_{3-y}]_3$ (A = Ba, Ca). The sodium chloride type layers are built up from double layers $[TlO]_\infty$, 2.2 Å apart, like in the $Tl_2Ba_2CaCu_2O_8$ structure, surrounded with $[A'O]_\infty$ layers, containing mainly barium, similar to the $[BaO]_\infty$ layers observed for $Tl_2Ba_2CaCu_2O_8$. In the triple perovskite layers the oxygen sites corresponding to the basal planes of the CuO_6 octahedra, i.e. parallel to (001) are fully occupied as well as the oxygen sites located at the same level as the barium ions. On the other hand, the anionic sites located at the same level as the calcium ions were found to be only 50 % occupied. Thus, such an oxygen distribution could be interpreted as resulting from the existence of simple layers of CuO_5 pyramids and double layers of CuO_5 pyramids and CuO_6 octahedra interleaved with calcium ions (Fig. 7b). This structure is closely related to that of $Tl_2Ba_2CaCu_2O_8$: it exhibits similar [BaO] and

9

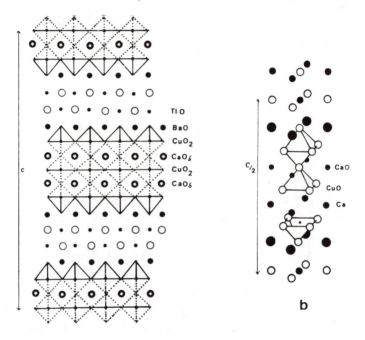

TlO
BaO
CuO$_2$
CaO$_6$
CuO$_2$
CaO$_6$

CaO
CuO
Ca

a

b

FIG. 7. a) Structural model of the 125K-supercon-
ductor Tl$_{2-x}$Ba$_2$Ca$_2$Cu$_3$O$_{10+\delta}$ and limit model to explain the presence of oxygen atoms in the calcium
planes (b).

[TlO]$_\infty$ layers, and differs from this latter oxide by
the introduction of one [CuO$_2$]$_\infty$ layer and additional oxygen between two pyramidal layers. Although
rather short, the distance between the copper planes, close to 3.6 Å, is much longer than that observed for Tl$_2$Ba$_2$CaCu$_2$O$_8$, so that the additional oxygen can be intercalated at the same level as the
calcium ions contrary to this latter oxide. The
deviation from stoichiometry deduced from the refinements of the structure, corresponding to the

10

formulation $Tl_{1.8}Ba_2Ca_2Cu_3O_{10+\delta}$ can be considered as significant, concerning the high $Cu(III)$ content of this oxide, in agreement with its high critical temperature and oxygen non-stoichiometry.

Very recently, we have isolated a new member of the thallium series, $TlBa_2Ca_2Cu_3O_{10-x}$, which exhibits a critical temperature of 120K and about 45 % diamagnetism[17]. A structural model of this phase
was established from X-ray powder data. It consists of triple oxygen-deficient perovskite layers similar to those observed in the "2223", and of double distorted rock salt type layers involving only one single thallium layer according to the sequence "BaO-TlO-BaO". Preliminary studies by high resolution electron microscopy give evidence of extended defects, and suggest the possibility of generation of new members in this series of intergrowths, according to the general formulation $[AO]_n$ $[A'CuO_{3-y}]_{n'}$ (A = Ba, Tl ; A' = Ba, Ca), where the integral numbers n and n' correspond to the multiplicity of the rock-salt type and perovskite - type layers respectively. The synthesis of higher n' members should allow the role of the thickness of

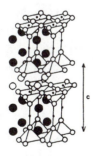

$Y Ba_2Cu_3O_6$

FIG. 8. Structure of the oxide $YBa_2Cu_3O_6$.

the oxygen -deficient perovskite layer in the increase of Tc to be explained. But more information
is also necessary concerning the oxygen content in
order to determine the Cu(III) content in those
oxides.

The mixed-valence of copper involving a partial
oxidation of Cu(II) into Cu(III)* is absolutely
necessary for superconductivity. Thus the oxygen
non-stoichiometry will play an important role in
the superconductivity of these oxides. $YBa_2Cu_3O_{7-\delta}$
gives a good example of the dramatic influence of
the oxygen content upon the critical temperature of
the orthorhombic phase. Although all the results do
not coincide from one author to the other, due to
the method of synthesis and also to the accuracy,
it is well established that Tc decreases drastical-
ly from 92K for $\delta \approx 0$, to 22K for $\delta < 0.6$, value
beyond which superconductivity disappears, as shown
for example by the results obtained by Tarascon et
al.[3], Monod et al.[18b], Cava et al.[19] and Tokumoto
et al.[20]. It is worth pointing out that for $\delta \geq$
0.50, the sole consideration of electroneutrality
would involve for copper a mean valence ranging
between 2 ($\delta = 0.50$) and 1.67 ($\delta = 1$). Thus this
part of the domain should not exhibit any superconductivity owing to the absence of the mixed-valence
Cu(III)-Cu(II). However,we have to take into consideration the coordination of copper. From the chemical point of view it is indeed well known that

_____

* Cu(III) represents a formal valence which can be
also understood as $Cu^{II}-O^-$ according to the results of Bianconi et al.[2b].

Cu(II) or Cu(III) can only take coordinations greater than 3, whereas Cu(I) takes only the linear
twofold coordination in oxides.

A model based on the continuous transition from the
YBa$_2$Cu$_3$O$_7$ structure (Fig. 1b) to the YBa$_2$Cu$_3$O$_6$
structure (Fig. 8) can be proposed. In order to
respect the coordination of copper, the different
compositions corresponding to $0 < \delta < 1$ will be
described as disordered intergrowths of the "O$_7$"
and "O$_6$" structures. In this model the pyramidal
layers [CuO$_{2.5}$] remain untouched, whereas [CuO$_2$]$_\infty$
rows of CuO$_4$ square planar groups, alternate with
[CuO]$_\infty$ rows of Cu^{I}O$_2$ groups, either in the same
layer (Fig. 9a) or from one layer to the next (Fig.
9b). Longitudinal intergrowths of CuO$_4$ square planar groups and Cu^{I}O$_2$ groups within a same chain can
also be considered (Fig. 9c), but such a mechanism
implies the creation of copper vacancies at the
junction of the chains, in order to avoid abnormal
coordination of copper ; in this latter case
chains must be long enough to avoid high concentration of copper vacancies in the structure. This

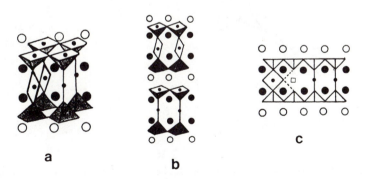

FIG. 9. Hypothetical models of "localized" intergrowths between the "O$_7$" and "O$_6$" structures.

structural model which is characterized by a partial disproportionation of Cu(II) into Cu(III) and Cu(I) leads to the formulation $YBa_2[Cu_2^{II}Cu^{III}O_7]_{1-\delta}[Cu_2^{II}Cu^I O_6]_\delta$. The examination of this formulation shows that the Cu(III) content is much greater than that deduced from the chemical analysis. Fig. 10a shows that, using this formulation, T_C increases approximately continuously as the ratio $[Cu(III)/Cu_{total}]$ (or $[Cu^{II}-O^-]/[Cu-O]_{total}$) increases for Monod's data, whereas the plateaus observed by different authors could be due to an ordering of the oxygen vacancies which is not yet really elucidated. This structural model explains well the presence of superconductivity for $\delta \geq 0.50$, by the existence of a mixed valence. Moreover it is supported by the observations by X-ray absorption of noticeable amounts of Cu(I) in all the domain $0 \leq \delta \leq 1$ simultaneously with Cu $3d^9 \underline{L}$ (Baudelet et al.[18d] ; Oyanagi et al[21]).

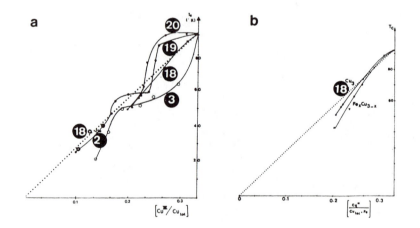

FIG. 10. a) $YBa_2 Cu_3 O_{7-\delta}$: T_C vs $[Cu(III)]/Cu_{TOT}]$. b) $YBa_2 Cu_{3-x} Fe_x O_7$: T_C vs $Cu[Cu(III)/Cu(Fe)]_{TOT}$. (numbers correspond to the references).

OXYGEN AND DEFECTS IN HIGH-T$_C$ SUPERCONDUCTORS

This prominent influence of the [Cu(III)/Cu$_{TOT}$] ratio upon the T$_C$ value is confirmed by the study of the oxides La$_{2-x}$Sr$_x$CuO$_{4-y}$. In this case no disproportionation of Cu(II) into Cu(III) and Cu(I) needs to be considered, since the oxygen deficiency does not introduce for copper a coordination smaller than four. From the data obtained by Kang et al.[18b] it can be established that there is no discontinuity in the evolution of T$_C$ versus Cu(III)/Cu(II) ratio (Fig. 10a). In the same way pure orthorhombic La$_{1.95}$CuO$_4$ prepared under high oxygen pressure (Tournier et al[2b]), which exhibits bulk superconductivity below 37K corresponds to a [Cu(III)/ Cu$_{TOT}$] ratio close to that observed for La$_{1.85}$Sr$_{0.15}$CuO$_4$ in agreement with the similar values of T$_C$ for these two compounds (see Fig.10a).

EXTENDED DEFECTS

The orthorhombic superconductor YBa$_2$Cu$_3$O$_{7-\delta}$ has been extensively studied by electron microscopy by several staffs simultaneously. A complete review of those results cannot be exposed here ; numerous details and references can be obtained from Ref. 22.

We would only like to draw attention to the fact that, besides the problems of microtwinning, numerous defects have been observed, according to the methods of the preparation of the powder samples or ceramics, which involve non-stoichiometry, especially on oxygen. An interesting point, which must be pointed out, concerns the particular study of YBa$_2$Cu$_3$O$_{6.6}$[22] for which different samples show, sometimes, doubling of the "b" parameter, sometimes

15

modulation of the contrast, corresponding to a "local" doubling of the "a" parameter. Such results, which correspond to "local" intergrowth of "O_7" and "O_6" structures, are in agreement with our model of disproportionation of Cu(II) into Cu(III) and Cu(I), described above.

SUBSTITUTION IN $YBa_2Cu_3O_{7-\delta}$

Many studies have been carried out in these last few months, in order to determine the influence of doping upon the superconducting properties of these phases (Tarascon et al.[3], Tsurumi et al.[24]). This confirms the very weak influence of the interleaved rare-earth ions upon superconductivity of these materials. On the other hand, Takita[23] has shown that T_C decreases dramatically as barium is partially replaced by rare-earth ions in the oxides $Ln_{1+x}Ba_{2-x}Cu_3O_{7-\delta}$ (Ln = La, Nd, Sm, Eu). This variation is accompanied by a decrease of the orthorhombic distortion, leading to a tetragonal symmetry around x = 0.20 for Nd which composition remains superconductive (T_C = 60K). The lack of data concerning the oxygen content does not allow one to decide here whether the decrease of T_C is due to the introduction of rare earth ions in the barium sites, or to a simple decrease of the [Cu(III)/Cu$_{TOT}$] ratio. In a rather similar way, $LaBa_2Cu_3O_{7-\delta}$ was found to have a smaller T_C, ranging from 60K (Nakai et al.[25]) to 75K (Michel et al.[2a]). The symmetry of this phase, which was reported initially as tetragonal by X-ray diffraction, was, in fact, found to be orthorhombic by electron microscopy (Hervieu et al.[2b]).

16

OXYGEN AND DEFECTS IN HIGH-T$_C$ SUPERCONDUCTORS

Several studies of the replacement of copper by different metallic elements have shown that only a progressive decrease of T$_c$ was observed, even for magnetic ions such as iron (see for instance Maeno et al.[3] and Oseroff et al.[3]). Here again, the lack of data concerning the oxygen content makes the chemical interpretation of those results difficult. However, more complete results obtained for the oxides YBa$_2$Cu$_3$O$_{7-\delta}$[26], have shown that the T$_c$ of those oxides decreases continuously as x increases from 92K for x = 0 to about 37K for x = 0.36. The recent structure determination of the oxide YBa$_2$Cu$_{2.85}$Fe$_{0.15}$O$_7$ by Roth et al.[27] shows definitely that this oxide is a superconductor (T$_c$ = 70K) in spite of its tetragonal symmetry, and that the iron is distributed in a preferential way between the pyramidal sites (0.05 Fe) and the oxygen deficient octahedra (0.1 Fe). The systematic study of those oxides by Deslandes et al.[28] confirms the superconductive character of the tetragonal form for x ranging from 0.15 (T$_c$ = 70K) to 0.35 (T$_c$ = 30K) and shows from the analysis of oxygen that δ remains close to 0.00 ± 0.02. The Mossbauer study of those compounds shows that the Fe(IV) is present, but in small amounts, compared to Fe(III). From these results, the curve of T$_c$ vs the ratio [Cu(III)]/ [(Cu+Fe)$_{TOT}$] can be plotted, assuming the approximate formula YBa$_2$[Cu^{II}Cu$_{1-x}$^{III}Fe^{II}$_x$]O$_7$ (Fig. 10b). It can be seen that this curve is near to that established for the pure oxide YBa$_2$Cu$_3$O$_{7-\delta}$ using Monod's data. The small difference between the two curves may partly be due to the presence of Fe(IV), which was not taken into consideration in our calculation. These results confirm the prominent role of the [Cu(III)/Cu$_{TOT}$] ratio in the superconductivity of these oxides, and suggests that the magnetic properties of iron are not so baneful as it could be expected from previous studies of classical superconductors.

Note added in proof : during this meeting at Tuscaloosa we were informed that the structure of the thallium "2223" had been recently determined from a single crystal X-ray diffraction study by Torardi et al.[29] ; the results obtained by these authors agree with our structural model. In the same way we learned that $TlBa_2Cu_3O_{10-x}$ was observed by Parkin et al.[30] but with $T_C = 110K$ instead of 120K and a superconducting factor of 10 to 20 % ; these differences are easily explained by the fact that these authors could not isolate a single phase for this composition.

REFERENCES

1. J.B. Bednorz and K.A. Muller, Z. Phys., B64, 189 (1986).
2. Caen "High Tc superconductors","blue booklets"
   a) Before 1987 and from January to May 1987,
   b) From May 1987 to August 1987.
3. Novel superconductivity Eds. W.V.Z. Krezin (Plenum Press, N.Y. 1987). Proceedings of the International Workshop on Novel Mechanism of Superconductivity, June 1987, Berkeley.
4. C. Michel, M. Hervieu, M.M. Borel, A. Grandin F. Deslandes, J. Provost and B. Raveau, Z. Phys., B68, 421 (1987).
5. H. Maeda, Y. Tanaka, M. Fukutomi and T. Asano Jpn J. Appl. Phys., in press.

6. R.M. Hazen, C.T. Prewitt, R.J. Angel, N.L. Ross, Fingen, C.G. Hadidiacos, D.R. Veblen, P.J. Heany, P.H. Hor, R.L. Meng, Y.Y. Sun, Y.K. Wang, Y.Y. Xue, Z.J. Huang, L. Gao, J. Bechtold and C.W. Chu. Phys. Rev. Lett., under press.

7. C. Politis, Appl. Phys., A45, 261 (1988).

8. M.A. Subramagnan, C.C. Torardi, J.C. Calabrese J. Gopalakrishnan, K.J. Morissey, T.R. Askew, R.B. Flippen, V. Chowdhry and A.W. Sleight, Sciences, under press.

9. J.M. Tarascon, Y. Lepage, P. Barboux, B.G. Bagley, L.H. Greene, W.R. McKinnon, G.W. Hull, M. Giroud and D.H. Hwang, Phys. Rev. B, under press.

10. M. Hervieu, C. Michel, B. Domengès, Y. Laligant, A. Lebail, G. Ferey and B. Raveau, Modern Physics Let. B, 2(1), 491 (1988).

11. Z.Z. Sheng, A.M. Herman, A. El Ali, C. Almasan, J. Estrada and T. Datta, Phys. Rev. Lett., submitted, Nature, submitted.

12. R.M. Hazen, L.W. Finger, R.J. Angel, C.T. Prewitt and C.G. Hadidiacos, Phys. Rev. Lett., submitted.

13. S. Parkin, V.Y. Lee, E.M. Engler, A.I. Nazzal, T.C. Huang, G. Gorman, R. Savoy and R. Beyers, Phys. Rev. Lett., submitted.

14. A. Maignan, C. Michel, M. Hervieu, C. Martin, D. Groult and B. Raveau, Modern Phys. Lett., in press.

15. M. Hervieu, C. Michel, A. Maignan, C. Martin and B. Raveau, J. Solid State Chem., to be published.

16. M.A. Subramanian, J.C. Calabrese, C.C. Torardi
    J. Gopalakrishnan, T.R. Askew, R.B. Flippen,
    K.J. Morrissey, U. Chowdhry and A.W. Sleight,
    Nature, 332 (1988).
17. C. Martin, C. Michel, A. Maignan, M. Hervieu
    and B. Raveau, C.R. Acad. Sci., under press.
18. Orsay high Tc superconductors preprints
    a) Vol. 1, April 1987 ; d) Vol. 4, Oct. 1987
    b) Vol. 2, June 1987 ; e) Vol. 3, Nov. 1987
    c) Vol. 3, July 1987 ; f) Vol. 6, Dec. 1987
19. R.J. Cava, B. Battlog, C.H. Chen, E.A. Riet-
    man, S.M. Zahurak, D. Werdea, Nature, 329, 423
    (1987).
20. M. Tokumoto, M. Ihara, T. Matsubara, M. Hira-
    bayashi, N. Terada, H. Oyanagi, K. Murata and
    Y. Kimura, Jpn J. Appl. Phys., 26, 1566 (1987)
21. H. Oyanagi, H. Ihara, T. Matsubara, M. Takumo-
    to, T. Matsuhita, M. Hirabayashi, K. Murata,
    Jpn J. Appl. Phys., 26, 1561 (1987).
22. B. Raveau, C. Michel, M. Hervieu and J. Pro-
    vost, Proceeding of the high Tc superconductor
    Conference, Interlaken, March 1988.
23. K. Takita, 9th Winter Meeting on Low Tempera-
    ture Physics High Tc superconductors, January
    1988.
24. S. Tsurumi, M. Hikita, T. Iwata, K. Semba and
    S. Kurihara, Jpn J. Appl. Phys., 26, L856
    (1987).
25. I. Nakai, K. Imai, T. Kawashima and R. Yoshi-
    zaki, Jpn. J. Appl. Phys., 26, L1244 (1987).
26. T.J. Kistenmacher, W.A. Bryden, J.S. Morgan,
    K. Moojani, Y.N. Du, Z.K. Qiu, H. Tang and
    J.C. Walker, Phys. Rev., B36, 8877 (1987).
27. G. Roth, B. Renker, G. Heger, V. Caignaert,
    M. Hervieu and B. Raveau, . Z. Phys., sub-
    mitted.

28. F. Deslandes, C. Michel, M. Hervieu, G. Heger
and B. Raveau, International Meeting of High
Tc superconductors. Schloss Mauterndorf,
Austria, February 7-11, 1988.

29. C.C. Torardi, M.A. Subramanian, J.C. Calabrese
J. Gopalakrishnan, K.J. Morrissey, T.R. Askev,
R.B. Flipper, U. Chowdry and A.W. Sleight,
Science, in press.

30. S.P. Parkin, V.Y. Lee, A.I. Nazzal, R. Savoy
and R. Beyers, submitted to Phys. Rev. Lett.

RECENT DEVELOPMENTS IN HIGH-TEMPERATURE
SUPERCONDUCTIVITY AT THE UNIVERSITY OF ALABAMA IN
HUNTSVILLE

M.K. WU*, B.H. LOO+, P.N. PETERS** and C.Y. Huang***
+Departments of Physics and Chemistry+, University
of Alabama-Huntsville, Huntsville, Alabama 35899; **
Marshall Space Flight Center, Space Science Laboratory,
Huntsville, Alabama 35812; *** Lockheed, Palo Alto,
California 94088

ABSTRACT

Superconducting 123 films were fabricated using the green
semiconducting 211 phase as a substrate.   High - temperature
processing was found to partially convert the green 211
phase oxide to the 123 phase. High-T_c superconductivity was
observed in Bi-Sr-Cu-O and Y-Sr-Cu-O systems prepared using
the same heat treatment process.   High-temperature processing presents an alternative synthetic route in the search
for new high-T_c superconductors.

1. INTRODUCTION

The discovery of superconductivity in the La-Ba-Cu-O system
(referred to as the "214" phase) in the 30K range by Bednorz
and Müller in 1986 [1] stimulated much interest among scien-
tists of many disciplines. Results of applying high pressure
[2], or replacing Ba^{2+} with other ions [3], which raised the
superconducting transition temperature T_c to above 50K,
suggested that higher-T_c compounds might be made by substituting the appropriate ion into this cuprate oxide system.
This supposition led us to the discovery of superconductivity at 90K in multiphase $Y_{1.2}Ba_{0.8}CuO_{4-y}$ [4].   Within weeks
of our discovery, the superconducting phase was identified
to be the black $YBa_2Cu_3O_7$ phase (referred to as the "123"
phase) and the accompanying green phase was shown to be the
semiconducting Y_2BaCuO_5 phase (referred to as the "211"
phase) [5].
     Immediately after the discovery, we found that Bi-Sr-
Cu-O, when prepared under extreme conditions, exhibited an
anomaly [6] indicative of superconductivity with an onset

temperature at about 60 K. However, an equilibrium phase was found to have a T_c of only 15 K.

While making the first 90K Y-Ba-Cu-O material, we observed that samples reacted at a temperature higher than 950°C, but for a comparatively short time, exhibited sharper superconducting transitions [7]. We also learned that an extremely careful heat treatment is required to prepare a single-phase bulk polycrystalline 123 compound. These results suggested that non-equilibrium processing may be thermodynamically favorable to the formation of the superconducting phase in the cuprate oxides. In fact, using high-temperature processing, we have successfully prepared the high temperature superconductors $BiSrCuO_{4-y}$ [8] and $YSrCuO_{4-y}$ [9]. In addition, an unusual magnetic suspension and enhanced critical current density was observed in a 123 and AgO composite [10]. In this paper we report on recent advances made in our laboratories.

2. EXPERIMENTAL

The compounds used as film substrates and in high-temperature processing were prepared in the following manner. Appropriate amounts of metal oxides were mixed, pressed into pellets, heated at 950°C for 12 hours, and then quenched to room temperature. Annealing procedures depended on the particular study being conducted, and are described in detail in appropriate sections. Electrical resistivity measurements were made with the conventional 4-probe technique. The DC magnetic moment measurements were made with a SQUID at the National Magnet Laboratory at MIT. A standard 4-probe using pulse current was used to determine the critical current density at zero field. Structural and phase determinations were made by x-ray diffraction and Raman microprobe analysis.

3. RESULTS AND DISCUSSIONS

3.1 Superconducting Thin Films

An RF sputtering technique was used to fabricate thin films. The 123 target material was first annealed in oxygen at 950°C for six hours, then slow-cooled in the furnace. Its superconducting onset temperature was about 96K. The 211 substrate material was prepared by heating and regrinding the oxides several times to ensure complete reaction. The homogeneity of the 211 phase was determined using a Raman probe [11].

The initially sputtered film was about 1 micron thick and was semiconducting. X-ray diffraction of the film showed single phase tetragonal 123 structure. It became superconducting after oxygen annealing at $900^{\circ}C$ for 30 minutes and furnace-cooling. Figure 1 shows the resistance of the annealed film as a function of temperature. The superconducting transition was at 93K and sharp. The critical current density, determined by 4-probe I-V measurement, is approximately 350 A/cm^2. This relatively low critical current can be attributed to the substrate materials which, were prepared with the conventional powder sintering technique. A compact substrate or even a single crystal 211 should give better films. The advantage of using a 211 substrate is that the transition temperature of the annealed sputtered film did not degrade. The relative ease of making the 211 substrate and the simple preparation conditions also make it a more attractive process An experiment using hot pressed 211 material as substrate is currently underway.

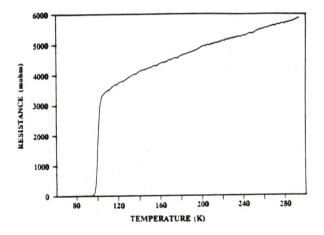

Figure 1: Electrical resistance of the annealed 123 film as a function of temperature.

3.2 High Temperature Processing

3.2.1 Superconductivity in Bi-Sr-Cu-O

Early in April 1987 [6], we observed a resistance anomaly at

60K in the Bi-Sr-Cu-O system prepared at 900°C with 30 minutes sintering time. However, a longer processing time at 850°C showed a complete superconducting transition at 15 K. After the success of using high-temperature processing to convert the green 211 phase to black 123 phase [8], we used the high-temperature processing technique to reexamine the system [8]. A material with nominal composition of BiSr-CuO_{4-y} was prepared from Bi_2O_3, SrO, and CuO, pressed into a pellet, and heated to 800o-850°C for 12 hours. The sample was quenched, reground, and annealed in oxygen for two hours at 1200°C and slow-cooled in the furnace. The samples melted during heating, but needle-like crystals were found inside the melt. Figure 2 shows the temperature dependence of the magnetic moment, m, for a sample (18 mg) cooled in fields of 10 and 100 G. Clearly the moment deviates around

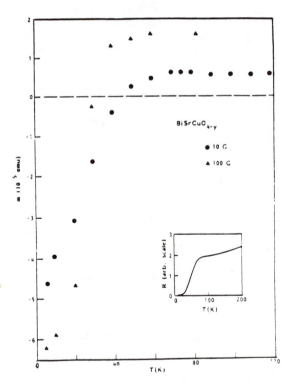

Figure 2: Magnetic moments of $BiSrCuO_{4-y}$ at 10 G and 100 G. The Inset is the temperature dependence of R.

60 K, suggesting a superconducting onset at 60 K. This onset temperature is in agreement with the onset of the resistance measurement, as shown in the inset of Figure 2. Figure 3 shows the field dependence when the sample was cooled in zero magnetic field. At low fields, m is linear in H, but starts to deviate at higher fields, thus defining H_{c1}. The H_{c1} value at 7 K is 20 G. As shown in Fig. 2, m is positive above T_c, indicating the presence of magnetic moments, which are presumably caused by the presence of Cu^{2+} from unknown phases. These moments might be the reason that m crosses zero at relatively low field values, as shown in Fig. 3. X-ray and Raman microprobe analysis (Fig. 4) of this compound differed from both those of the 214 and 123 phases, indicating the possible existence of a new phase. Structure determination of this system is currently underway.

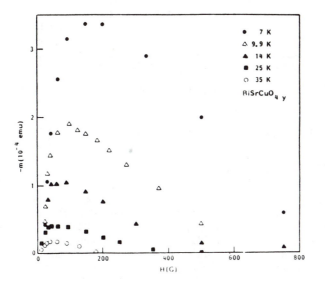

Figure 3: Field dependence of the magnetic moment for $BiSrCuO_{4-y}$ at various temperatures.

Two important conclusions can be drawn from this study. First, ions other than rare earth metals can form new copper oxide compounds that still exhibit high-temperature superconductivity. Second, high-temperature processing may be favorable to the formation of certain high-temperature copper oxide systems.

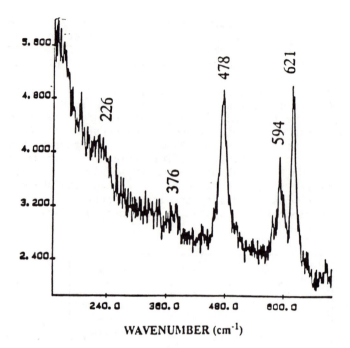

Figure 4: Raman spectrum of $BiSrCuO_{4-y}$.

3.2.2. Superconductivity in Y-Sr-Cu-O

Complete replacement of the Ba^{+2} ions with the smaller alkaline earth metal ions has not produced high-T_c superconductors. Thus far only $Y_{0.3}Sr_{0.7}CuO_{3-y}$, with a T_c of about 40 K, has been reported [12]. Based on the successful results on Bi-Sr-Cu-O, we decided to reexamine the Y-Sr-Cu-O [9] system using similar high-temperature processing conditions.

A sample with a nominal composition $YSrCuO_{4-y}$ was prepared by mixing appropriate amounts of Y_2O_3, SrO, and CuO. The mixture was ground and pressed into pellets, heated to $1300^{\circ}C$ for 2h, and then quenched to RT. The material was then reground, repressed, reheated to $1200^{\circ}C$ for 6 h in O_2, and then slowly cooled to RT. High-purity alumina crucibles were used in the sample preparation.

28

The temperature dependence of the resistance for an Y-Sr-Cu-O sample is displayed in Figure 5. The superconducting onset is about 80 K, which is consistent with that of the magnetic moment measurement (sample mass is 25 mg; external field 10 G) shown in the inset. A linear temperature dependence of R before the onset of superconductivity was observed. The resistance curve of our sample does not indicate the presence of a second superconducting phase. However, magnetic moment data indicate one superconducting transition at 80 K and another at 40 K. Figure 6 shows the field dependence of magnetic moment at various temperatures. These curves are similar to those for $BiSrCuO_{4-y}$.

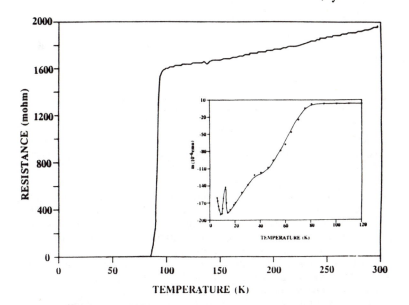

Figure 5: Resistance as a function of temperature of Y-$SrCuO_{4-y}$. Inset illustrates its magnetic moment at 10 G.

The 80 K superconducting phase did not form when our samples were prepared at a lower temperature. This indicates that the 80-K phase is thermodynamically stable only at higher temperatures. This is consistent with the conversion of the green 211 phase to the corresponding 123 phase with good superconducting characteristics by processing at 1300°C.

In addition to the two superconducting transitions, the magnetic moment curve shows an anomaly at low temperatures.

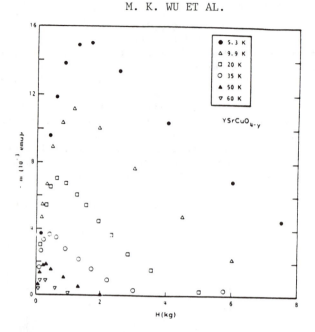

Figure 6: Field dependence of magnetic moment at various temperatures.

Figure 7 shows details of the low-temperature data for the sample cooled in 20, 50, 100, 200, 1000 and 2500 G. The magnetic moment is diamagnetic for $T < T_c$, but there is a sharp reduction in the diamagnetic moment at 14 K, indicating a sharp drop in superconductivity. It is not likely that this is caused by the onset of magnetic ordering, because the moment above T_c is small. It is not presently clear whether there is a structural phase transition at this temperature, or whether an unidentified phase is present.

3.3 $YBa_2Cu_3O_7$ and AgO Composite

AgO decomposes when it is heated above $200^\circ C$ and does not react with high-T_c superconductors. Therefore, a $YBa_2Cu_3O_7$ composite can be formed by mixing $YBa_2Cu_3O_7$ and AgO powders, then processing them under certain conditions. Because of the high conductivity of the silver component, the $YBa_2Cu_3O_7$-AgO composite may exhibit better superconducting characteristics than the plain $YBa_2Cu_3O_7$ material.

Some of the typical resistance curves of the $YBa_2Cu_3O_7$-

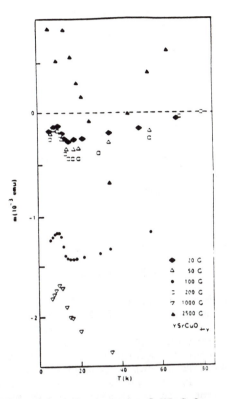

Figure 7: Magnetic moment of YSrCuO$_{4-y}$ at different fields.

AgO composites are shown in Figure 8. The room temperature conductivity increases as the silver content increases. Measurements of critical current density showed that, J_c 250 to 350 A/cm^2 at 77K, were about two orders of magnitude higher than for our undoped samples at the same temperature. The composites were more ductile and malleable. We have also observed a stable suspension [10] of the samples in the divergent field of magnets. While superconducting, they can be suspended above (the normally observed case) or below the magnet. When suspended below an attractive force, equal to the weight plus any repulsion, is provided by flux pinning. A total weight of 2.07 times the sample weight was success-fully lifted from a 77K surface with a magnet having a maximum field of 3.3 kG and maximum gradient of 2.6 kG/cm.

A number of conclusions can be drawn from our observa-tions. The decrease in normal-state resistivity and the

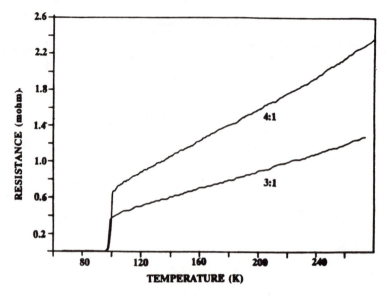

Figure 8: Electrical resistance of $YBa_2Cu_3O_7$-AgO composites. Ratios are weight of 123 to AgO reagent.

increase in critical current density with the addition of silver oxide suggest a lowering of contact resistances between grains. Near identical T_c's suggest that grain interiors experienced little change. We model the magnetic properties and suspension results in terms of many weak-link current loops, where the weak links are associated with the inter-grain contacts. When a sample is brought into the gradient field of a magnet, the increasing field induces diamagnetic currents in all loops. When a critical current is exceeded in some loop, flux enters that loop, dropping the current momentarily and trapping the flux. Bringing the sample closer to the magnet results in additional flux trapping in weaker loops, and further diamagnetic current growth in stronger loops. The induced diamagnetic currents generate repulsive forces in the gradient field. If at any point in this process the sample moves away from the magnet, the changes in currents and field reverse. Loops with no trapped flux continue to cause repulsion, which is decreasing. However, the currents in the loops with trapped flux will decrease and eventually reverse direction, attempting to prevent flux flow out of the loops; this generates an attractive force in the gradient field. Doping with silver oxide enhances the critical currents of inter-grain contacts, providing stronger pinning forces that can provide the suspension of the sample beneath the magnet.

The M-H hysteresis [13] loop for a sample (weight ratio 3:1) measured at 77K is shown in Figure 9 to a maximum field H_{max} = 0.9 kOe. The measurement was taken after the sample was cooled in zero field, and was independent of the sweep rate between 0.03 and 0.4 kOe/min. The critical current density, estimated from the magnetization at 0.1 kOe is 3.3×10^{3} A/cm^{2}, based on Bean's critical state model [14]. This value of J_c is an order of magnitude larger than that measured from transport properties. The residual magnetization is 3 emu/g, and is not sensitive to a magnetic field larger than 0.5 kOe. A slight decrease of the residual magnetization with time was observed up to 10 minutes.

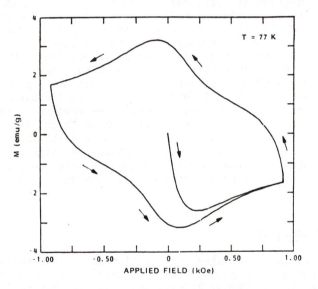

Figure 9: M-H hysteresis loop for 123 + AgO (3:1 weight ratio) at 77 K.

A crucial feature of the data as shown in Fig. 9 is the development of a positive magnetization when the sweep direction is reversed at H_{max}. In fact, M becomes positive at a high field, which is only about 0.1 kOe below H_{max}. The change in sign of the magnetization when H is decreased only slightly, after reaching H_{max}, gives rise to the unusual magnetic suspension of a superconductor in the field gradient below a permanent magnet. As described in reference

33

6: when a strong permanent magnet (P) was moved toward one
of these 123-AgO superconductors (S) at 77 K, S was first
repelled by P. This is consistent with Fig. 9, which shows
that M < 0 (diamagnetic) during the initial increase of H
from 0 to H_{max}. However, if P was first moved close to S
(field at S $>$ 0.3 kOe), then pulled away from S the field at
S was reduced such that M became positive, resulting in S
being attracted to P. As a result of this attraction, S
stayed stably suspended below P.

5. SUMMARY

The studies presented here indicate that proper processing procedures are critical to the formation of high-temperature copper oxide superconductors. Superconducting 123
films can be fabricated using the green 211 phase as a
substrate. The superconducting characteristics of these
films, in terms of superconducting transition width, are
better than the characteristics found when other oxide compounds are used as substrates. A compact or single crystal
211 phase will be desirable as a substrate for high-quality
thin films. A new high-T_c copper oxide compound with non-
rare earth elements was prepared using high temperature
processing. High-temperature processing presents an alternative synthetic route in the search for new high-T_c superconductors. A $YBa_2Cu_3O_7$-AgO composite with improved electrical conductivity and strong flux pinning was also prepared.

ACKNOWLEDGEMENTS

The work at the University of Alabama in Huntsville was
supported by NASA grants NCC8-2, NAGW-812, and NAG8-089, and
by NSF Alabama EPSCoR RII-8610669, and at Marshall Space
Flight Center by the Center Director's Discretionary Fund.
The authors would like to thank J.R. Ashburn, C.A. Higgins,
I.E. Theodoulou, W.E. Carswell, D.H. Burns, A. Ibrahim, T.
Rolin, and R.C. Sisk for technical assistance.

REFERENCES

1. J.G. Bednorz and K.A. Müller, Z. Phys.B64, 189 (1986).
2. C.W. Chu, P.H. Hor, R.L. Meng, L. Gao, Z.J. Huang, and
    K.Q. Wang, Phys. Rev. Lett. 58, 405 (1987).

3. R.J. Cava, R.B. van Dover, B. Batlogg, and E.A. Reitman, Phys. Rev. Lett. 58, 408 (1987).
4. M.K. Wu, J.R. Ashburn, C.J. Torng, P.H. Hor, R.L. Meng, L. Gao, Z.J. Huang, Y.Q. Wang, and C.W. Chu, Phys. Rev. Lett. 58, 908 (1987).
5. R.J. Cava, Batlogg, R.B. van Dover, D.W. Murphy, S. Sunshine T. Siegrist, J.R. Remeika, E.A. Reitman, S. Zahurak, and G.P. Espinosa, Phys. Rev. Lett. 58, 1676 (1987).
6. Reported at International Conference on Cryogenic Materials, Chicago, May 1987, and also at Japan-U.S. Symposium on High Temperature Superconductivity, Tokyo, October 1987.
7. M.K. Wu, J.R. Ashburn, and C.J. Torng, unpublished results.
8. M.K. Wu, J.R. Ashburn, C. Higgins, B.H. Loo, D.H. Burns, A. Ibrahim, T. Rolin, P.N. Peters, R.C. Sisk, and C.Y. Huang, Appl. Phys. Lett. June 1, 1988.
9. M.K. Wu, J.R. Ashburn, C. Higgins, B.H. Loo, D.H. Burns, A. Ibrahim, T. Rolin, F.Z. Chien, and C.Y. Huang, Phys. Rev. B, June, 1988.
10. Peters, P.N.; Sisk, R.C.; Urban, E.W.; Huang, C.Y.; and Wu, M.K., Appl. Phys. Lett. (June 20, 1988).
11. B.H. Loo, M.K. Wu, D.H. Burns, A. Ibrahim, C. Jenkins, T. Rolin, Y.G. Lee, D.O. Frazier, and F. Adar, in "High Temperature Superconducting Materials-Preparation, Properties, and Processing," ed. W.E. Hatfield, 1988.
12. Y. Mei, S.M. Green, C. Jiang, and H.L. Luo, in "Novel Superconductivity", edited by S.A. Wolf and V.Z. Kresin (plenum, New York, 1987), p. 1041.
13. C.Y. Huang, E.J. McNiff, P.N. Peters, B. Schwartz, Y. Shapira, and M.K. Wu, to be published.
14. Bean, C.P. Phys. Rev. Lett., 1962, 8, 250.

OXYGEN DEFICIENCY AND ITS EFFECT ON NORMAL AND SUPERCONDUCTING PROPERTIES OF $Ba_2LnCu_3O_{7-\delta}$ (Ln=RARE EARTHS)

TETSUYA HASEGAWA, KOHJI KISHIO, KOICHI
KITAZAWA and KAZUO FUEKI*
Department of Industrial Chemistry, University of Tokyo,
Bunkyo-ku, Tokyo 113, Japan

Abstract Normal and superconducting properties of $Ba_2LnCu_3O_{7-\delta}$ (Ln = rare earths) are examined from the viewpoint of defect chemistry by measurements of the oxygen nonstoichiometry, the electrical conductivity and the chemical diffusion coefficient.

INTRODUCTION

In high-T_c oxide superconductors, the oxygen deficiency is known to be one of the major factors that determines their normal and superconducting properties.[1] It is especially a key parameter in $Ba_2LnCu_3O_{7-\delta}$ - type superconductors (Ln = rare earths), because the Ba, Ln, Cu mole fractions are fixed, in contrast to the $(La, M)_2CuO_{4-\delta}$ (M = alkaline earth element) system, in which the mole fraction of M substantially affects its superconductivity. By increasing the oxygen deficiency δ in $Ba_2LnCu_3O_{7-\delta}$, the superconducting transition temperature (T_c) is decreased,[2,3] accompanied by an orthorhombic-tetragonal phase transition. This is why one needs either low-temperature annealing, or a slow cooling process after sintering, to achieve a higher T_c.

In this paper, we report the equilibrium value of the oxygen deficiency in $Ba_2LnCu_3O_{7-\delta}$ (Ln = Y, Pr, Nd, Sm, Gd), systematically

---

* Present Address: Department of Industrial Chemistry, Science
University of Tokyo, Noda 278, Japan.

investigated at various temperatures and oxygen partial pressures. The obtained data are treated thermodynamically, in order to clarify the equilibrium value of the oxygen deficiency δ. The electrical conductivity at high temperature was also investigated, in relation with the normal-state properties of $Ba_2LnCu_3O_{7-\delta}$. The relative difference of conductivities, particularly the orthorhombic-tetragonal phase transition temperature (which appears as the inflection point in the conductivity-temperature curve), is discussed in terms of the ionic radius of the rare earth ion. Furthermore, preliminary results of chemical diffusion coefficient measurements for the Y system will be described.

OXYGEN NONSTOICHIOMETRY

Nonstoichiometry measurements were performed with a thermogravimetric microbalance under an O_2/Ar controlled atmosphere.[4] Specimens were prepared by a conventional powder mixing of Ln_2O_3, $BaCO_3$ and CuO in ethanol. The absolute oxygen contents was determined by iodometric titrations for the specimens which were annealed in 1 atm O_2 at 300°C.

The oxygen deficiences (δ) of $Ba_2LnCu_3O_{7-\delta}$ as a function of temperature and oxygen partial pressure, $p(O_2)$, are summarized in Figure 1. No discontinuity was seen in any δ-$p(O_2)$ curve, indicating that the orthorhombic-tetragonal phase transition is of second order. All figures, as a whole, showed a quite similar tendency, namely, each system can reach a large oxygen deficiency (δ) up to 0.8-0.9, beyond which it decomposes. On the other hand, the oxygen content (7-δ) does not exceed 7, at which Cu-O linear chain is complete. In fact, the maximum oxygen contents of the five systems investigated (Ln = Y, Nd, Sm, Gd, Pr) are identical (6.92-6.95) within experimental error

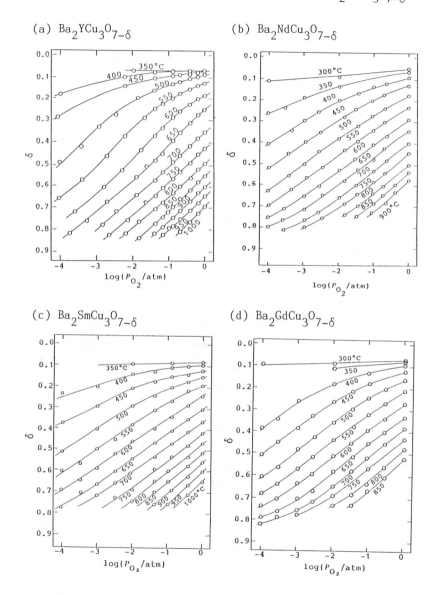

FIGURE 1. Oxygen nonstoichiometry of $Ba_2LnCu_3O_{7-\delta}$ (Ln=Y, Nd, Sm, Gd) as a function of temperature and Po_2.

(±0.02). In the Pr system, the Pr ion is assumed to have valence 4+, as described later, so that $Ba_2PrCu_3O_7$ seems to be stably formed. However, its oxygen content showed the same limited value.

The present results are in contrast with the effect of Cu site doping on oxygen deficiency.[5] The oxygen content increases by substituting Fe or Co for Cu, and exceeds 7 at about 10% substitution, although the Cu-O linear chain or the orthorhombic structure are destroyed by the impurity doping. It is obvious that the maximum oxygen content of 6.9 is not essential for the orthorhombic structure and its superconductivity. However, the experimental facts mentioned above may suggest that the orthorhombic 1-2-3 compounds with $\delta = 0$ have a structural instability, which is supported by the observation that the oxygen content and the crystal structure in the Y system do not change by placing the sample under high O_2 pressure (up to 40 atm.)

Some differences in oxygen deficiency between Ln-substituted systems can be seen at lower temperatures. Figure 2 compares the δ-T relations at $p(O_2)=1$ atm. As can be seen, the δ values of Ln = Nd, Sm, Gd have almost the same temperature dependence, while the Y system show a smaller oxygen deficiency in the temperature region 400-700°C.[4,6-8] This result means in practice that the Nd, Sm and Gd systems need a lower-temperature annealing to incorporate enough oxygen, compared to the Y system.

We proceed into a thermodynamic treatment of the defect equilibrium.[9] When considering the solution of oxygen molecules into a solid lattice, the partial molar enthalpy $\Delta H(O_2)$ and partial molar entropy $\Delta H(O_2)$ are expressed as follows:

$$\Delta H(O_2) = \frac{\partial(R \ln p(O_2))}{\partial(1/T)} \tag{1}$$

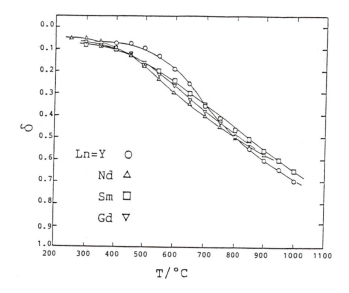

FIGURE 2. Oxygen nonstoichiometry δ of $Ba_2LnCu_3O_{7-\delta}$ (Ln = Y, Nd, Sm, Gd) at $p(O_2)$ = 1 atm.

$$\Delta S(O_2) = -\frac{\partial(RT \ln p(O_2))}{\partial T} \qquad (2)$$

Thus, $\Delta H(O_2)$ and $\Delta S(O_2)$ can be deduced from the slopes of $R\ln p(O_2)$ vs $1/T$ and $RT\ln p(O_2)$ vs T plots, respectively. Both plots for the Nd system are shown in Figures 3 and 4; the resulting thermodynamic parameters are given in Figure 5. Both $\Delta H(O_2)$ and $\Delta S(O_2)$ curves indicate a discontinuity around δ = 0.3, which is attributed to the orthorhombic-tetragonal phase transition. Above δ = 0.3, $\Delta H(O_2)$ increases monotonically with δ. This behavior of $\Delta H(O_2)$

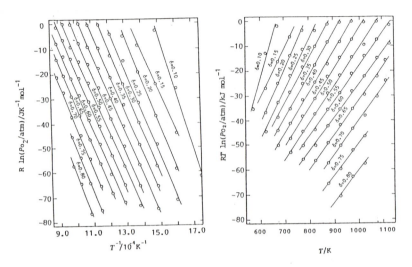

Figure 3. Rlnp(O₂) vs 1/T plot
for $Ba_2NdCu_3O_{7-\delta}$

Figure 4. RTlnp(O₂) vs T plot
for $Ba_2NdCu_3O_{7-\delta}$.

in the Nd system differs from that in the Y system, which shows another discontinuity at $\delta = 0.5$.[9]

The $\Delta S(O_2)$-δ relationship in Figure 5 is dominated by the configurational entropy of oxygen vacancies:

$$\Delta S(O_2) = 2[\Delta S(O_2)^0 + \Delta S_{conf}] \qquad (3)$$

Therefore, the defect equilibrium may be understood by assuming an appropriate defect model to fit the configuration entropy to the experimental $\Delta S(O_2)$. We considered the following models, taking account of two inequivalent Cu sites, Cu(I) and Cu(II), and electron holes:

(1) Holes are delocalized.

(2) Holes are localized on Cu(I) site.

42

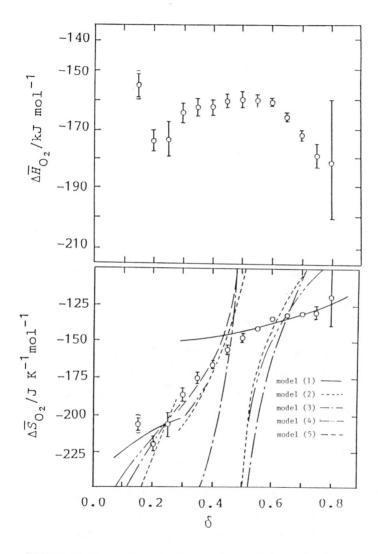

FIGURE 5. Partial molar enthalpy and entropy for
oxygen dissolution in $Ba_2NdCu_3O_{7-\delta}$.

(3) Holes are localized on Cu(II) site.

(4) Holes are localized on Cu(I) and Cu(II).

(5) Holes are localized on O or (Cu-O).

In all models, it is assumed that the oxygen vacancy exists only on the Cu(I) plane, based on the neutron diffraction study.[10]

Theoretical $\Delta S(O_2)$-δ curves are fitted to the experimental one by adjusting $\Delta S(O_2)^0$. The results of the curve-fitting are also included in Figure 5. For the region $\delta = 0.5 - 0.8$, where the crystal structure is tetragonal, the delocalized electron hole model (model (1)) is consistent with the experimental result, rather than the localized electron hole models. Below $\delta = 0.5$, on the contrary, all theoretical curves fail to show good agreement. For the region $\delta < 0.4$, it seems that model (5), assuming O^- instead of Cu^{3+}, describes well the observed behavior of $\Delta S(O_2)$. However, the $\Delta S(O_2)$-δ curves, predicted by all localized electron hole models, indicate a discontinuity at $\delta = 0.5$, due to the disproportionation reaction $2Cu^{2+} \longrightarrow Cu^+ + Cu^{3+}$, and hence the defect models proposed here may require some essential improvement, such as the involvement of mixed valence in both Cu(I) and Cu(II) sites, and the incorporation of both localized and delocalized holes.

ELECTRICAL CONDUCTIVITY

The electrical conductivites σ of $Ba_2LnCu_3O_{7-\delta}$ (Ln = Y, La, Pr, Nd, Sm, Eu, Gd, Dy, Ho, Er, Tm, Yb) were measured over a temperature range from 300 to 900°C in 1 atm O_2 atmosphere by a dc four-probe technique.[11] The observed temperature dependence of the conductivities is presented in Figure 6. Since the sintering densities were slightly different, depending on the compounds, the temperature dependences of σ in Figure 6 are considered to be essentially the same for Ln=Nd through Yb and Y. On the other hand, the conductivities of the La and Pr systems exhibited weaker temperature dependences, and, in particular, the La system showed significantly lower conductivity at room temperature.

The decrease in σ with increasing temperature, seen in all curves in Figure 6, is mainly due to the fact that oxygen deficiency increases with temperature. Introduction of oxygen deficiency reduces the carrier concentration, resulting in the decrease in σ.

The distinctive behavior of σ at high temperature, in comparison with the nonstoichiometry data of Figure 1, is that the inflection points, corresponding to the orthorhombic-tetragonal phase transition,[12,13] are clearly seen, except for the La system. Figure 7 is a replot of the data of Figure 6, by normalizing σ to the value at the inflection point. The phase transition temperature (T_{O-T}) appears to be correlated with atomic number, except for Y and Pr. However, if T_{O-T} is plotted against the ionic radius of the trivalent rare earth ion, as shown in Figure 8, the T_{O-T} values lie on the single master curve, except for the Pr system. For the T_{O-T} of the La system in Figure 8, the data from an X-ray diffraction study were used.[14] Thus, it is concluded that the ionic radius of the rare earth element is a major factor in determining the phase transition temperature.

In order to investigate the exceptional behavior of the Pr system, we performed resistivity and Hall coefficient measurements at low temperature. The observed ρ and R_H values are given in Table I, along with carrier concentration and Hall mobility μ_H estimated from ρ and R_H. As the temperature decreases, μ_H increases, while the carrier concentration p is significantly reduced. This implies the presence of an energy gap, which may result from the higher valence of Pr ion (possibly 4^+).

45

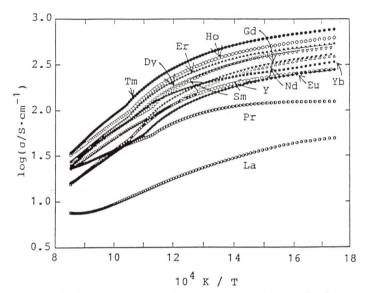

FIGURE 6. Electrical conductivities of $Ba_2LnCu_3O_{7-\delta}$ at $Po_2=1$ atm

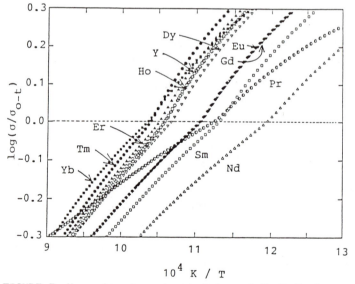

FIGURE 7. Normalized conductivities of $Ba_2LnCu_3O_{7-\delta}$ at $Po_2=1$ atm.

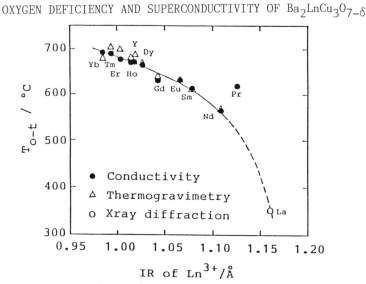

FIGURE 8. Correlation plot between orthorhombic-tetragonal transition temperature T_{O-T} and ionic radius (IR) of Ln^{3+}.

Table I. Low-temperature resistivity and Hall coefficient data of $Ba_2PrCu_3O_{6.94}$.

T(K)	$\rho(\Omega cm)$	$R_H(cm^3C^{-1})$	$p(cm^{-3})$	$\mu_H(cm^2V^{-1}s^{-1})$
11.8	2.79×10^2	5.1×10^2	1.2×10^{16}	1.8
190.0	1.42×10^{-1}	1.7×10^{-2}	3.8×10^{20}	1.2×10^{-1}

CHEMICAL DIFFUSION COEFFICIENT

When an oxide specimen is equilibrated with O_2 atmosphere, and the oxygen partial pressure is suddenly changed, the weight of the specimen approaches the equilibrium value in a exponential manner. If

47

the specimen is an infinitely large thin plate, the weight change with time is given by the following equation:[15]

$$\frac{w(t)-w(\infty)}{w(0)-w(\infty)} = \frac{8}{\pi^2} \sum_{j=0}^{\infty} \frac{1}{(2j+1)^2} \exp[-(\frac{-(2j+1)\pi^2}{h})\tilde{D}t] \quad (4)$$

where $w(t)$ and $w(\infty)$ are the sample weight at time t and $t = \infty$, respectively, and $\tilde{D}$ is the chemical diffusion coefficient. Thus, $\tilde{D}$ can be estimated from the weight change curve by solving Eqn (4).

In this $\tilde{D}$ measurement, the specimen is required to be dense enough to satisfy the assumption in Eqn.(4). The sintering density ratio of $Ba_2YCu_3O_{7-\delta}$ specimens, prepared by a conventional powder mixing, has been generally 70-90%, which is insuffcient in the present experiment. We succeeded in preparing a specimen with sintering density of 98% by a coprecipitation method with oxalic acid and ethanol. The details of the preparation procedure will be described elsewhere.

The $\tilde{D}$ measurement was made by using a thermogravimetric microbalance. Figure 9 shows a typical relaxation curve observed at T = 789 K. No hysteresis was observed in oxidizing and reducing cycles. By performing the same measurement on specimens with different thicknesses, the diffusion process was found to be diffusion-controlled. The corresponding $\tilde{D}t$ vs t plot is given in Figure 10.

The Arrhenius plot of the estimated $\tilde{D}$ is shown in Figure 11. In the orthorhombic region, represented by closed circles, the diffusion is quite slow, so that only a few data could be obtained. $\tilde{D}$ values, in the δ region investigated in the present study, ranged from 10^{-7} to 10^{-5} cm^2s^{-1}, which is in accordance with the $\tilde{D}$ values of typical perovskite-type oxides such as $(La_{1-x}Sr_x)CoO_{3-\delta}$.[16] The activation energy, evaluated from the slope of the Arrhenius plot in Figure 11, is about 115 kJ/mol, independent of δ, which is also consistent with those of perovskites.

48

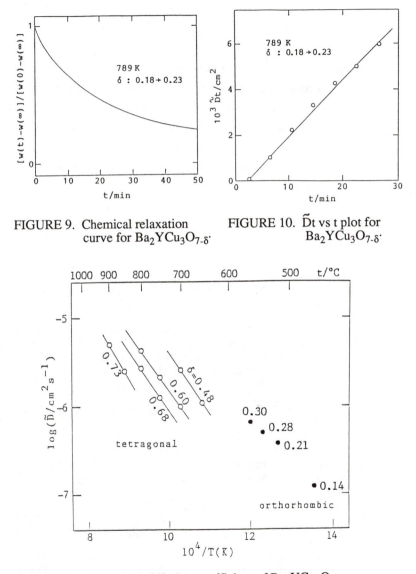

FIGURE 9. Chemical relaxation curve for $Ba_2YCu_3O_{7-\delta}$.

FIGURE 10. $\widetilde{D}t$ vs t plot for $Ba_2YCu_3O_{7-\delta}$.

FIGURE 11. Chemical diffusion coefficient of $Ba_2YCu_3O_{7-\delta}$.

T. HASEGAWA, K. KISHIO, K. KITAZAWA AND K. FUEKI

It should be noted here that, at the same temperature, $\tilde{D}$ decreases with increasing δ. Diffusion of oxygen is considered to occur only in the Cu(I) plane, and the distance between Ba-Ba planes becomes wider as oxygen vacancies are introduced. The present observation is apparently unreasonable, since the diffusion process seems to take place more rapidly when the distance from the adjacent plane is larger. Therefore, it seems that the interactions between oxygen atoms in the plane must be taken into consideration, along with the interaction between the planes, and, in that sense, the elucidation of the defect equilibrium is expected.

REFERENCES

1. K. Fueki, K. Kitazawa, K. Kishio and T. Hasegawa, ACS Symp. Ser. #351, Chapter 4, Eds. D. L. Nelson, M. S. Whittingham and T. F. George, Amer. Chem. Soc. (1987).
2. H. Takagi, S. Uchida, H. Iwabuchi, E. Eisaki, K. Kishio, K. Kitazawa, K. Fueki and S. Tanaka, Physica, 148B, 349 (1987).
3. R. J. Cava, B. Batlogg, C. H. Chen, E. A. Rietmann, S. M. Zahurak and D. Werder, Phys. Rev., B36, 5719 (1987).
4. K. Kishio, J. Shimoyama, T. Hasegawa, K. Kitazawa and K. Fueki, Jpn. J. Appl. Phys., 26, L1228 (1987).
5. J. M. Tarascon, P. Barboux, L. H. Greene, G. W. Hull and B. G. Bagley, Phys. Rev. B, in press.
6. E. Takayama-Muromachi, Y, Uchida, M. Ishii, T. Tanaka and K. Kato, Jpn. J. Appl. Phys., 26, L1156 (1987).
7. Y. Kubo, Y. Nakabayashi, J. Tabuchi, T. Yoshitake, A. Ochi, K. Utsumi, H. Igarashi and M. Yonezawa, Jpn. J. Appl. Phys., 26, L1888 (1987).
8. P. K. Gallagher, Adv. Cer. Mat., 2, 632 (1987).
9. K. Kishio, T. Hasegawa, J. Shimoyama, N. Ooba, K. Kitazawa and K. Fueki, Proc. Int. Conf. Sintering Sci. Techn., Tokyo, Oct. 16-19 (1987).
10. J. D. Jorgensen, M. A. Beno, D. G. Hinks, L. Soderholm, K. J. Volin, R. L. Hitterman, J. D. Grace, I. K. Schuller, C. U. Segre, K. Zhang and M. S. Kleefisch, Phys. Rev., B36, 3608 (1987).
11. S. Kambe, K. Kishio, N. Ooba, N. Sugii, K. Kitazawa and K. Fueki, Jpn. J. Appl. Phys. Supplement (1988).
12. P. P. Freitas and T. S. Plaskett, Phys. Rev., B36, 5723 (1987).

13. A. T. Fiory, M. Gurvitch, R. J. Cava and G. P. Espinosa, Phys. Rev., B36, 7262 (1987).
14. A. Maeda, T. Yabe, K. Uchinokura, M. Izumi and Tanaka, Jpn. J. Appl. Phys., 26, L1550 (1987).
15. K. Fueki, K. Kitazawa, K. Kishio and T. Hasegawa, Proc. Meeting High-T_c Superconductors, Schloss Mauterndorf, Austria, Feb. 7-11 (1988).
16. K. Fueki, J. Mizusaki, S. Yamauchi, T. Ishigaki, Y. Mima, Reactivity of Solids (Proc. 10th Int. Symp.), Elsevier, 339 (1985).

PREPARATION AND CHARACTERIZATION OF ISOTOPIC OXYGEN-ENRICHED YTTRIUM BARIUM COPPER OXIDE

KEVIN C. OTT, JAMES L. SMITH, ROBERT M. AIKIN, LUIS BERNARDEZ,[1] WILLIAM B. HUTCHINSON, ERIC J. PETERSON, EUGENE J. PETERSON, C. BALLARD PIERCE,[2] JAMES F. SMITH, JOE D. THOMPSON, AND JEFFREY O. WILLIS
Los Alamos National Laboratory, Los Alamos, New Mexico, USA

Abstract The preparation of labeled superconducting yttrium barium copper oxides from the labeled metal nitrates is described. The materials were characterized by a variety of physical techniques, and their superconducting properties were measured. Trends are seen in the structural parameters of the materials obtained by the nitrate route and also in isotopically enriched materials prepared via gas-phase exchange. These structural changes are accompanied by changes in superconducting properties. The ^{18}O-enriched yttrium barium copper oxide prepared via the nitrate route has a T_c depressed by 33 K to 59 K, whereas a ^{17}O sample has a T_c between the ^{18}O material and a similarly prepared ^{16}O sample.

Isotopic substitution of high-temperature superconductors has been the subject of intense interest to researchers intrigued with the role of electron-phonon interactions in the mechanism for superconductivity in these materials. By the old wisdom, T_c is generally thought to be a sensitive function of phonon frequency, i.e., the ionic mass, but coulomb correlations can dominate this small effect. If the electron-phonon interaction is significant in the high T_c phases, we still expect that T_c will be similarly influenced by the mass of the ions involved in critical phonon modes. In the yttrium barium copper oxide system, substituting the much heavier lanthanides at the yttrium site has had no significant effect on the superconducting properties.[3] Similarly, isotopic substitution of the barium and copper sites has not led to any observable changes in T_c.[4] Changing the mass of the oxide lattice in $YBa_2Cu_3O_{7-x}$ by partial substitution of ^{18}O for ^{16}O has led to the observation of a small decrease in T_c of <0.5 K upon isotopic oxygen substitution.[5-8] The technique for incorporating ^{18}O into the lattice of the "123" materials in these reports has been gas-phase exchange, which relies on all of the lattice oxide ions being mobile among the different sites on the time scale of the experiment. When we initiated our experiments, we did not expect complete exchange to occur during gas-phase exchange because of the chemical diversity of the sites that the oxide ions occupy. Assuming an oxygen stoichiometry of 7 in the superconducting phase, the lack of exchange in one site results in only 6/7 of the oxygen being exchanged. Our hypothesis was that much greater than 6/7 of the oxygen needs to be substituted to verify the presence or lack of a significant isotope effect. The earliest reports of isotopic substitution by gas-phase exchange claimed that from 65-90 at. % ^{18}O had been incorporated into the lattice by using various conditions of time and temperature of annealing under $^{18}O_2$. At only 65 at. % substitution, it is possible that two sites are completely unexchanged, whereas at 90 at. % enrichment, one site may remain completely unexchanged. Because most of the gas-phase exchange experiments reported in the literature used large excesses of >95 at. % ^{18}O dioxygen, one would expect levels of enrichment from gas-phase exchange to be greater than 90 at. %. The estimated levels of gas-phase exchange are approximately 90 at. % and lower, reflecting the fact that the exchanges did not reach equilibrium. To assess the oxygen isotope effect, it was necessary to substitute the lattice with >95 at. % ^{18}O, and to perform the substitution in such a manner that there would be no question regarding the homogeneity of isotopic substitution in all sites. At Los Alamos, we have examined the possibility of preparing the "123" phase from a variety of synthetic routes that incorporate the isotopes directly into the lattice upon synthesis. Los Alamos is examining the preparation[9] of thin film superconductors via coevaporation of yttrium, copper, and BaF_2 followed by annealing in flowing $^{18}O_2/H_2{}^{18}O$ as a method to incorporate ^{18}O directly into the lattice.[10] We also are examining the synthesis of the "123" phase via oxidation of a

55

2:1 mixture of the binary alloys BaCu and YCu in $^{18}O_2$ (Ref. 11). Results of these experiments will be published elsewhere. In this paper, we describe our results on materials prepared from labeled metal nitrates and $^{18}O_2$ gas-phase exchange.

Because the "123" phase may be prepared by decomposing mixtures of yttrium, barium, and copper nitrates in an oxygen atmosphere, we synthesized the isotopically enriched metal nitrates using highly ^{18}O-enriched nitric acid. We also prepared samples highly enriched in ^{17}O and ^{16}O for comparison.

All "123" samples and the yttrium powder used as a starting material were handled under purified argon in a glove box or on a Schlenk[12] line using standard Schlenk techniques to avoid contamination with ^{16}O atmospheric constituents. Isotopically enriched nitric acid was prepared by reacting $^{15}N^{18}O$ with $H_2^{18}O$ in the presence of a slight excess of $^{18}O_2$ (Ref. 13). The $H^{15}N^{18}O_3$ prepared was isolated and purified by vacuum distillation. Similarly, ^{17}O-enriched nitric acid was prepared from $N^{16}O$, ^{17}O-enriched water, and ^{17}O-enriched oxygen. The isotopic composition of the $HN^{18}O_3$ (in atomic %) was ^{16}O-4%, ^{17}O-0.4%, and ^{18}O-95.6%. The $HN^{17}O_3$ used was ^{16}O-26.8%, ^{17}O-40.5%, and ^{18}O-32.7%.

Samples of isotopically enriched $YBa_2Cu_3{}^*O_{7-x}$ (asterisk denotes isotopic enrichment of ^{16}O, ^{17}O, or ^{18}O) were prepared by dissolving yttrium and copper metal and $BaC^{16}O_3$ in an excess of 50%

$HN^{*}O_3$. All samples were prepared in an identical
fashion. In a typical preparation, 177.2 mg
yttrium powder, (99.999%, Alfa), 389.4 mg copper
wire (99.999%, Aldrich), and 759.1 mg $BaCO_3$
(99.999%, Aldrich) are dissolved under a counter-
flow of argon in 5 cm^3 of cold (0°C) $HN^{*}O_3$.
Vigorous evolution of H_2 and $N^{*}O_x$ occurs during
the addition of the metals. Carbon dioxide
effervesces rapidly during addition of the $BaCO_3$,
and $BaN^{*}O_3$ precipitates. A transfer arm and
receiver are attached to the reaction flask and
are flushed with argon. The suspension is frozen
at 77K, and the apparatus is evacuated. Excess
$HN^{*}O_3$ and $H_2^{*}O$ are removed from the metal
nitrates by vacuum distillation at room temperature, which yields a light-blue powder of the
hydrated, mixed metal nitrates. The $HN^{*}O_3$ recov-
ered from the ^{18}O synthesis contained 94.2 at. %
^{18}O, 1.3 at. % ^{17}O, and 4.5 at. % ^{16}O. The
isotopic content of the nitric acid may be
slightly degraded because of isotopic exchange
with $C^{16}O_2$ from the $BaCO_3$. We consider 94 at. %
to be the lower limit of ^{18}O incorporation into
the subsequently formed ^{18}O-enriched "123"
compound, which has been confirmed to be 95-96
at. % ^{18}O by laser ionization mass spectroscopy
(Lawrence Livermore National Laboratory) for a
sample of ^{18}O-labeled yttrium barium copper
oxide. The mixture of hydrated mixed metal
nitrates is then pyrolyzed under flowing argon at
450° to 500°C for 2 hours to yield a gray-black
material. The reaction flask is transferred into

the glove box, and the residue is scraped into a mortar and ground with a pestle to a fine dark-gray powder. The powder is then pressed to a loose pellet (intentionally porous to facilitate oxygen diffusion). The pellet is placed in a boat (platinum or quartz), isolated in a quartz tube, and removed from the drybox. The apparatus is placed into a clamshell furnace, attached to a vacuum line, and evacuated.

After backfilling the vessel with the appropriate isotopically enriched oxygen, a constant flow of <1 cm^3 *O$_2$/min at 580 torr (atmospheric pressure in Los Alamos) is established. The reaction vessel is isolated from the atmosphere by two cold traps (-78°C) and a bubbler in series. The initial heating cycle is as follows: the temperature is ramped from 30° to 200°C in 5 min and held for 5 min, is ramped to 650° in 30 min and held at this temperature for 1 hour. N*O$_x$ evolution begins at approximately 350°C and appears to be complete after 1 hour at 650°C. The temperature is then raised over a period of 30 min to 960°C, held for 4 hours, then cooled under a constant flow of *O$_2$ to 100°C over a period of 200 min. The apparatus is isolated from the atmosphere, and the sample is returned to the glove box, where the sample is reground and pelletized, and the heating process under *O$_2$ is repeated with the 1-hour hold time at 650°C eliminated. Two samples each of the ^{16}O- and ^{17}O-enriched "123" and three samples of the ^{18}O-enriched "123" superconductors were prepared with this annealing procedure. In addition, some ^{18}O samples were annealed a third time at 700°C for 1.5 hour, cooled at 2°/min to 500°C, held at this temperature for 30 min, and then cooled to room temperature in 100 min. Samples were examined by x-ray diffraction (XRD), transmission electron microscopy (TEM), electron microprobe, scanning electron microscopy (SEM), and chemical analysis; in addition, the superconducting properties of the samples were measured.

Examination of the powder XRD results indicates that the materials prepared by the nitrate route often contain small amounts of $Y_2BaCu^*O_5$ ("211") and some Cu^*O. This result is confirmed by SEM/electron microprobe examination. TEM found no observable differences in morphology of the various isotopically enriched materials. No significant differences were found between the nitrate-derived or typical oxide/carbonate-derived ^{16}O-enriched materials, so it appears that the mode of synthesis does not influence the properties of the materials. Elemental analyses were performed to obtain the bulk composition (atomic emission spectroscopy). Analyses of single grains were determined by electron microprobe using both energy- and wavelength-dispersive spectrometers. Results of these determinations are shown in Table 1. There appears to be a trend in the stoichiometry of individual grains in samples containing the heavier isotopes prepared by the nitrate route. As the mass of the

TABLE 1   Analyses[a] of Isotopic Samples Prepared via the Nitrate Route

Isotope	Y	Ba	Cu
^{16}O	1.07 (1.08)	2.0	3.17 (3.01)
^{16}O	1.07 (1.07)	2.0	3.11 (3.03)
^{16}O[b]	– (1.04)	2.0	– (2.96)
^{17}O	0.98 (1.05)	2.0	3.14 (2.88)
^{17}O	1.07 (1.11)	2.0	3.08 (2.90)
^{18}O	0.97 (1.04)	2.0	2.82 (2.85)
^{18}O	– (1.03)	2.0	– (2.85)
^{18}O	1.02 (1.04)	2.0	3.05 (2.82)

[a]Bulk stoichiometry determined by atomic emission spectroscopy. Electron microprobe data from examination of 10-12 grains of the "123" phase are indicated in parentheses. All values normalized to Ba=2.
[b]Sample prepared via the carbonate/oxide route.

oxygen isotope increases, the amount of copper in single grains of "123" decreases, and the amount of $Cu^{*}O$ inclusions found in the bulk increase. These inclusions are not taken up into the lattice of the copper-deficient superconducting phase even upon extended firing in $^{18}O_2$. This suggests that a chemical isotope effect may occur during the decomposition of the nitrates that controls the copper stoichiometry of the resulting superconducting phase. One such ^{18}O-enriched sample, which was "deficient" in copper in the grains ($YBa_2Cu_{2.85}O_{7-x}$) but had a large number of

Cu^*O inclusions and a bulk stoichiometry of $YBa_2Cu_3^*O_{7-x}$, was fired in $^{16}O_2$. Subsequently, XRD and microprobe examination indicated that the "excess" CuO inclusions had been taken up into the lattice. Thus, there appears to be a subtle balance in the phase composition of the yttrium barium copper oxide system, which is dependent upon the oxygen isotope present in the lattice.

Examination of unit cell parameters obtained from high-resolution XRD indicates additional trends among the isotopically labeled material obtained via the nitrate route (Table 2). As the mass of the oxygen isotope substituted into the lattice increases, the a and b parameters increase significantly, and the c parameter decreases substantially. The unit cell volume remains fairly constant and at values consistent with complete oxygenation. These trends are not consistent with the trends observed in increasingly oxygen deficient ^{16}O samples, where the c

TABLE 2 XRD Data for Nitrate-Derived Y Ba Cu Oxides

Sample	a(Å)	b(Å)	c(Å)	vol(Å^3)	T_c
^{16}O	3.8189	3.8867	11.680(1)	173.37	93.5
^{17}O	3.8247	3.8885	11.674(2)	173.62	78
^{18}O	3.8304	3.8897	11.669(3)	173.86	59

Standard deviations for the a and b parameters are 0.0005 Å.

parameter and unit cell volume are observed to increase substantially upon removal of oxygen.[14-15] We have compared the nitrate-derived samples with gas-phase-exchanged samples that were prepared by repeated extended firing of ^{16}O-enriched "123" in [18]O_2. As the exchange proceeds and presumably as the amount of [18]O in the lattice increases, the a and b parameters increase, whereas the c parameter and unit cell volume remain at constant values indicative of complete oxygenation (Table 3). The increase in the a and b parameters, observed in the gas-phase-exchanged samples, is substantially less than the increase observed in the nitrate-derived materials. The above experiments indicate that incorporation of [18]O into the lattice by either

TABLE 3   XRD Data for [18]O_2 Gas-Phase-Exchanged Samples

Sample	a(Å)	b(Å)	c(Å)	vol(Å^3)	T_c
[16]O[a]	3.8177	3.8860	11.679(1)	173.26	93
[18]O[b]	3.8226	3.8879	11.679(1)	173.56	92.1
[18]O[c]	3.8238	3.8891	11.678(2)	173.66	90.4
[18]O[d]	3.8221	3.8879	11.674(2)	173.48	92.7

[a]Starting material.
[b]Two exchange cycles at 950°C, total time of 53 hours.
[c]Sample b, with two more cycles at 950°C, total time 111 hours.
[d]Sample c exchanged with [16]O_2, 950°C, 10 hours.

gas-phase exchange or preparation from the labeled nitrates subtly alters the structure. At this time, we do not know why there is such a large difference in the structural behavior of the ^{18}O-enriched gas-phase-exchanged samples and the nitrate-derived samples. We are attempting to prepare large samples of enriched materials for powder neutron diffraction studies of these phenomena.

The resistivity of the isotopically enriched nitrate-derived materials is shown in Figure 1. No corrections have been made for porosity. The onsets determined from magnetic susceptibility and ac resistivity measurements for the ^{18}O-, ^{17}O-, and ^{16}O-enriched materials (given as resistive midpoint and the 10-90% width in parentheses) are 59(5.5) K, 77(10) K, and 93.0(1.5) K, respectively. All samples examined had metallic normal state resistivity signatures, indicative of adequate oxygenation. The ^{17}O sample had an oxygen stoichiometry of 7.3 ± 0.4 with a combination of Rutherford back scattering and oxygen isotopic analysis. These observations indicate that the isotopically labeled materials are adequately oxygenated, consistent with the XRD data discussed above.

Gas-phase exchange of ^{16}O into the ^{18}O- and ^{17}O-enriched samples (4 hours at 960°C in flowing $^{16}O_2$ followed by a slow cool to room temperature) resulted in increased T_c's. The T_c of the ^{18}O nitrate-derived sample increased from 59 to 76 K (width of 7 K), whereas the T_c of the ^{17}O sample

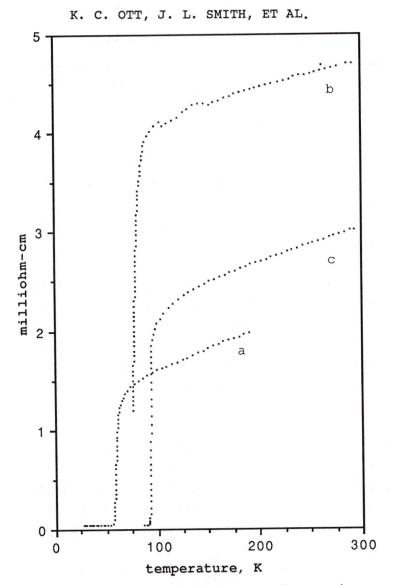

FIGURE 1. Resistivity versus temperature for isotopically enriched superconductors: (a) ^{18}O, (b) ^{17}O, (c) ^{16}O. ^{16}O- and ^{17}O-enriched samples mesured in a N_2 cryostat, hence the temperature range is limited to above 77 K. Porosity of the samples (by SEM) falls in the order b>c>a.

increased from 77 to 90.5 K (width of 1.7 K).
The ^{17}O-enriched sample, even after exchange with
$^{16}O_2$, has a T_c lying between the ^{18}O- and ^{16}O-
enriched materials, which suggests that the
depression in T_c is related to the mass of the
oxygen isotope. The shifts that we observe in
the nitrate-derived materials are surprisingly
larger than the shifts observed in samples prepared by gas-phase exchange. Because we also
observe surprisingly large differences in the
structural parameters of the isotopically labeled
nitrate-derived samples, the question is not how
the heavier isotopes decrease T_c, but rather, how
do the heavier isotopes affect the structure of
the superconducting phase? Since there is no
precedent for the preparation of highly enriched
perovskites in the literature, we have little
guidance as to what to expect structurally with
high levels of isotopic enrichment. However, it
is surprising to us that there is any structural
perturbation upon enrichment with the heavier
isotopes. Another point that we are addressing
is the difference in gas-phase exchanged samples
relative to the nitrate-derived samples.
Currently, we are carefully characterizing a
number of gas-phase-exchanged samples to determine the influence of copper deficiency and
degree of exchange upon the structural and superconducting properties of these materials. Early
indications are that copper deficiency in ^{16}O
yttrium barium copper oxide (stoichiometry

65

measured in individual grains by electron microprobe analysis) does not measurably effect T_c and
that there is only a slight perturbation of the a
lattice parameter. Gas-phase exchange of ^{18}O
into copper deficient ^{16}O samples is not observed
to affect T_c to any greater extent than exchange
into stoichiometric "123" samples. At this time,
it appears that copper deficiency in the nitrate-
derived samples does not explain the structural
differences, or the changes in T_c we observe in
materials prepared by the nitrate route.
Clearly, the unusual physical properties of the
isotopically enriched nitrate-derived yttrium
barium copper oxide superconductors is a manifes-
tation of the unusual chemistry of this system.

ACKNOWLEDGMENTS

The authors thank M. Goldblatt and T. E.
Walker for synthesizing the labeled nitric
acids, J. R. Fitzpatrick for providing the
isotopes and isotopic analyses of starting
materials, G. E. Bentley for analytical support,
R. B. Roof for assistance with XRD, the Los
Alamos Ion Beam Materials Laboratory for RBS
measurements, and M. Fluss of Lawrence Livermore
National Laboratory for assistance with the SIMS
and LIMS isotope ratios.

Work performed under the auspices of the
U.S. Department of Energy, Office of Basic Energy
Sciences, Division of Materials Science.

PREPARATION OF LABELED Y BA CU OXIDE

REFERENCES

1. Lawrence Livermore National Laboratory
2. Williams College
3. Z. Fisk, J. D. Thompson, E. Zirngiebl, J. L. Smith, and S. W. Cheong, Solid State Commun. 62, 643 (1987).
4. L. C. Bourne, A. Zettl, T. W. Barbee III, and Marvin L. Cohen, Phys Rev. B 36, 3990 (1987)
5. L. C. Bourne, M. F. Crommie, A. Zettl, H.-C. zur Loye, S. W. Keller, K. J. Leary, A. M. Stacey, K. J. Chang, M. L. Cohen, and D. E. Morris, Phys. Rev. Lett. 58, 2337 (1987).
6. B. Batlogg, R. J. Cava, A. Jayaraman, R. B. van Dover, G. A. Kourouklis, S. Sunshine, D. W. Murphy, L. W. Rupp, H. S. Chen, A. White,, K. T. Short, A. M. Mujsce, and E. A. Rietman, Phys Rev. Lett. 58, 2333 (1987).
7. K. J. Leary, H.-C. zur Loye, S. W. Keller, T. A. Faltens, W. K. Ham, J. N. Michaels, and A. M. Stacey, Phys Rev. Lett. 59, 1236 (1987).
8. H. Katayama-Yoshida, T. Hirooka, A. J. Mascarenhas, Y. Okabe, T. Takahashi, T. Sasaki, A. Ochiai, T. Suzuki, J. I. Pankove, T. Ciszek, and S. K. Deb, Jpn. J Appl. Phys. 26, L2085 (1987).
9. P. M. Mankiewich, J. H. Scofield, W. J. Skocpol, R. E. Howard, A. H. Dayem, and E. Good, Appl. Phys. Lett. 51, 1753 (1987).
10. C. J. Maggiore, M. A. Nastasi, and J. R. Tesmer, personal communication.
11. R. B. Schwarz, personal communication.
12. D. F. Shriver and M. A. Drezdon, "The Manipulation of Air Sensitive Compounds", Wiley Interscience, New York, 1987, pp. 30-31.
13. M. Goldblatt and T. E. Walker, manuscript in preparation.
14. R. J. Cava, B. Batlogg, C. H. Chen, E. A. Rietman, S. M. Zahurak, and D. Werder, Phys. Rev. B 36, 5719 (1987).
15. W. K. Kwok, G. W. Crabtree, A. Umezawa, B. W. Veal, J. D. Jorgensen, S. K. Malik, L. J. Nowicki, A. P. Paulikas, and L. Nunez. Phys. Rev. B 36, 106 (1988).

PARACONDUCTIVITY STUDIES ON $YBa_2Cu_3O_7$ SINGLE CRYSTALS

N. P. ONG, S. J. HAGEN, Z. Z. WANG, and T. W. JING

Joseph Henry Laboratories, Princeton University, NJ 08544

Abstract The paraconductivity within the CuO_2 planes has been measured in $YBa_2Cu_3O_7$ crystals. We find that with the proper subtraction procedure to remove the normal "background", the 2D Aslamasov–Larkin behavior is observed for T as high as 240 K. The crossover to the zero-resistance state is much more abrupt than predicted by the theory of Lawrence and Doniach. We interpret the disagreement as a failure of mean-field theory close to T_c.

Introduction

One of the interesting issues in the study of the superconducting behavior in the high-T_c oxides[1,2] is the question of the electronic anisotropy and the dimensionality of the fluctuation effects at temperatures T above the critical temperature T_c. Various studies have appeared of the anisotropy in the upper critical field[3-5] H_{c2} and in the normal-state resistivity[7-9]. Measurements of the paraconductivity, mostly on ceramic samples or thin film samples, have also been reported[10-13].

We have recently extended such measurements to single crystals of $YBa_2Cu_3O_7$ which display very sharp transitions above 90 K. These crystals were grown by a BaO-CuO flux technique at temperatures T near 980°C. Presynthesized $YBa_2Cu_3O_7$ powder was initially dissolved in the molten flux[8,14]. Slow cooling of the solution below the crystallisation point results in large crystals of $YBa_2Cu_3O_7$ (up to 2 x 2 x 0.08 mm^3.) As reported elsewhere[8,14] all as-grown samples exhibit a sharp transition (10% to 90% in 0.5 K) with a T_c between 90 and 93 K. The in-plane resistivity ρ_{ab} measured by the Montgomery technique varies from 120 to

leaving only the AL term. In some systems such as Pb thin films the MT term is not important, presumably because δ is intrinsically large.

In layered compounds such as $2H$-NbSe$_2$ and intercalated $2H$-TaS$_2$ the electronic anisotropy imposes an anisotropy in the superconducting parameters such as ξ and H$_{c2}$. At large ε, when the coherence length normal to the layers ξ_c is shorter than the interlayer separation d, the paraconductivity is 2D. As T$_c$ is approached, ξ_c grows until the system crosses over to 3D behavior ($\sigma' \sim 1/\sqrt{\varepsilon}$) near T$_c$. Retaining the AL terms only, Lawrence and Doniach[19] have used an anisotropic effective mass in the GL equations to derive the expression

$$\sigma_{LD}' = (e^2/16\hbar s) [1/\{ \varepsilon (1+ v(\varepsilon))\}^{1/2}] \tag{4}$$

where s is the interlayer separation and $v(\varepsilon) = (2 \xi_c/s)^2 = (2 \xi_c(0)/s)^2/\varepsilon$. In Eq. (4) a crossover from 2D to 3D behavior occurs at a temperature T$_c$(1 + v(0)). A similar 2D to 3D cross-over expression valid for the Maki-Thompson term has also been derived.

Paraconductivity in YBa$_2$Cu$_3$O$_7$

From the relatively low in-plane conductivity of YBa$_2$Cu$_3$O$_7$ (compared with Al) one expects that the paraconductivity should make a substantial correction to the normal state conductivity σ_N. This was in fact the case from early measurements performed on sintered YBa$_2$Cu$_3$O$_7$ by Tsuei et al[10]. However, compared with conventional superconductors there are three technical difficulties in analysing the data. First, the background σ_N is difficult to determine. This is especially problematical if the fluctuations are 2D in character, in which case the behavior $\sigma' \sim 1/\varepsilon$ at large ε predicted by Eqs. 2-4 is difficult to disentangle from a normal state resistivity which is linear in T with a positive intercept ρ_0, i.e.

$$\sigma_N = 1/(aT + \rho_0), \qquad (\rho_0 > 0) \tag{5}$$

To leading order in 1/T, Eqs. 2-5 all vary as 1/T. Secondly, we do not have the convenience of using an intense field to suppress the superconducting fluctuations. For reasons that are not understood, the paraconductivity is little affected by fields up to 20 T. Thirdly, the transitions are fairly broad, so that T_c cannot be determined with high precision. This precludes an accurate study of the divergence of σ' very near T_c.

In sintered samples the usual procedure has been to assume that the σ_N is accurately determined by fitting the total observed conductivity σ_T above ~200 K to a straight line. The paraconductivity is then obtained from $\sigma' = \sigma_T - \sigma_{N1}$ where σ_{N1} is an extrapolation of the fit to lower T. However, this procedure fails to isolate the fluctuation contribution in 2D as discussed above. In a preliminary report[14] we adopted this procedure and found that, in a log-log plot of σ' vs. ε, the paraconductivity falls off much faster than any reasonable power law at large ε. A good fit of the data for σ' (obtained with the incorrect subtraction procedure) is given by

$$\sigma' = -G \log \varepsilon + B, \qquad (G, B > 0) \qquad (6)$$

As indicated, the logarithmic fit is valid only in a temperature range from 1 K to 30 or 40 K above T_c. Other groups[20,21] have also reported a similar fit to the data. This unusual behavior led us to re-examine the arbitrary subtraction procedure used.

To measure the paraconductivity in single crystals, we have extended our previous technique[8]. Contacts (with resistances < 10 Ω) are made to each crystal using 25 μm Au wires tipped with pure In. At each T the temperature is stabilised to $\pm$ 0.1 K, and two resistances R_1 and R_2 are measured using the 4-probe lock-in technique operating near 27 Hz. The separate components ρ_{ab} and ρ_c are computed using Montgomery's technique. Near T_c the data points are separated by 0.1 K, while at higher T the separation approaches 1 K. The out-of-plane fluctuations are not discussed here because of the larger uncertainty of the background. However, they are roughly ten times smaller than the in-plane paraconductivity very near T_c. Because of sample inhomogeneities, what appears to be a "sharp" transition at T_c can be less than ideal for such studies. A good indicator of inhomogeneity is the derivative $d\rho_{ab}/dT$ of the resistivity. In **Fig. 2** we show for 4 samples the derivatives plotted vs. T.

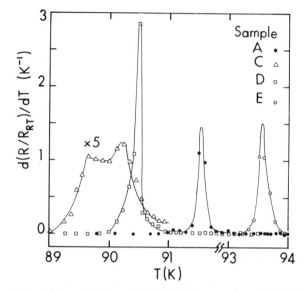

Figure 2 Plot of $(d\rho_{ab}/dT) / \rho_{ab}(300K)$ *vs* T for samples A, C, D, E. The widths (measured at half height) are less than 0.2 K for A, D and E.

Samples with ΔT (defined as the width of the derivative peak at half-height) larger than 1 K typically show sub-structure in the peak (e.g. Sample C). It is clearly impossible in such samples to select a T_c with any degree of confidence. Most of our analyses are performed on the three samples A, D and E which have exceptionally narrow ΔT (under 0.1 to 0.2 K). Even so, some uncertainty in determining T_c remains.

The reanalysis of our data proceeds as follows. If the fluctuations are essentially 2D for large ε, the total conductivity is given by

$$\sigma_T = (aT + \rho_0)^{-1} + C/(T- T_0) \tag{7}.$$

In the second term we have distinguished the temperature T_0 at which divergence occurs from the observed critical temperature T_c. To first order in $1/T$ the two terms in Eq. 7 are additive. However, higher-order terms differ in sign. The first term in Eq. 7 saturates as T decreases, whereas the second term diverges. Instead of studying σ_T, we multiply σ_T by T to remove

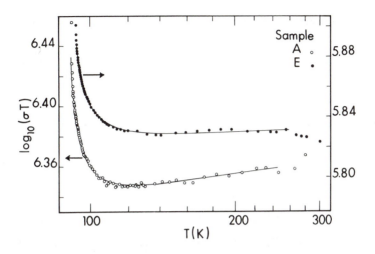

Figure 3 Fit of the total conductance σ_T to Eq. 7 shown on log-log scale. (The product $\sigma_T T$ is plotted to emphasize the curvature near 100 K.) The solid line is the best fit to Eq. 7.

the leading term in 1/T. By plotting the product $\sigma_T T$ vs. T **(Fig. 3)** we find that the countervailing trends produce a shallow minimum at

$$T_m = \sqrt{(\rho_0 C T_0)} \{1 + \sqrt{(T_0 / C\rho_0)}\} / \{1 - a\sqrt{(CT_0/\rho_0)}\} \qquad (8)$$

provided $\rho_0 > a^2 C\, T_0$. By fitting the shallow well to the corresponding expression from Eq. 7, we then determine the parameters a, ρ_0, C and T_0. Although this procedure does not guarantee that the fit is unique, we feel that it is more sensible than the arbitrary subtraction used in our earlier report. In particular, the plot of $\sigma_T T$ isolates the strength of the diverging term C somewhat more unambiguously. For e.g., we note that if the second term in Eq. 7 is replaced by fluctuations of the 3D form $C/\sqrt{(T - T_0)}$, then $\sigma_T T$ should diverge as $\sqrt{T}$ at large T. This case can be definitely excluded from our data. After a and ρ_0 are determined from the fit, σ is computed from $\sigma_T - 1/(aT + \rho_0)$ at all T.

We find that all data from four samples can be accurately fitted using this procedure if we allow T_0 to differ significantly (a few K) from T_c **(Fig. 3.)** See **Table 1** for the parameters. In Samples A and B the

75

Sample	a (m Ω cm/K)	ρ_0(m Ω cm)	C [(Ω cm)$^{-1}$]	I_0 (K)	I_c (K)
A	0.423	5.34	267	87.8	91.5
B	0.647	0.809	283	87.0	90.6
C	0.592	0.850	71	88.8	89.9
D	0.554	16.3	50	88.7	90.4
E	1.46	7.87	69	89.3	93.5

Table 1 Parameters obtained from fit to Eq. 7 (see text) using data 1 K above T_c. The theoretical value of C is 262 (Ω cm)$^{-1}$.

strength C [267 and 283 (Ω cm)$^{-1}$ respectively] is found to be close to the theoretical value $e^2/16\hbar s$ (= 262 (Ω cm)$^{-1}$) if we use for the average distance s between CuO_2 planes half the unit cell dimension along **c** (d = 5.7 Å.) In the other three samples, however, C is smaller. The close agreement in C with the theoretical AL value (in Samples A and B) encouraged us to

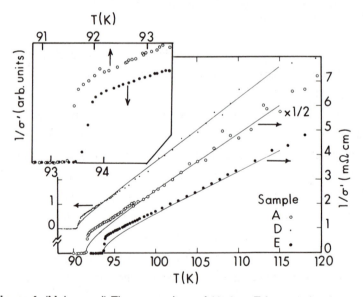

Figure 4 (Main panel) The comparison of 1/ σ' vs T for samples A, D, and E with the best fit (solid lines) using the Lawrence–Doniach model. The disagreement is particularly serious near T_c for Samples A, and E. (Inset) The behavior of 1/ σ' near T_c for Samples A and E. Data points are spaced 0.1 K apart near T_c.

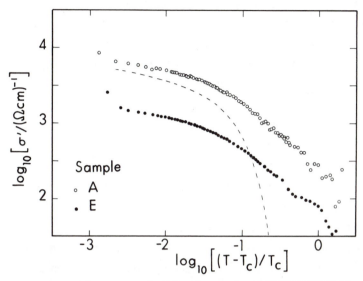

Figure 5 Plot of the paraconductivity σ' vs. reduced temperature ε on log-log scale for samples A and E. The data points are calculated from σ$_T$(T) - (aT + ρ$_0$)$^{-1}$ where (aT + ρ$_0$)$^{-1}$ is derived from the fit to Eq. 7. An alternate procedure used to determine σ$_N$ by fitting σ$_T$ to 1/(aT + ρ$_0$) in the temperature range 200-300 K) generates data for Sample A which fall on the broken line.

further investigate the behavior closer to T$_c$. In the traditional Parks plot of 1/σ' vs. T the cross-over from the MT regime to the AL regime is conveniently displayed since 1/σ$_{AL}$' is linear in T. In the LD model, however, 1/σ$_{LD}$' varies as √{ ε (1+ ν(ε))}.

We find that the transition to the zero-resistance state, when it occurs, is much sharper than described by the LD effective mass theory. In **Fig. 4** we show the plot of 1/σ' vs. T for Samples A,D and E. It is clear that the data are consistently above the LD curve,which is much more rounded near T$_c$. (The data agree with the LD curve at large ε since both vary as 1/ε.) An alternate way to display the data is by a log-log plot of σ' vs. ε (**Fig. 5.**) We find that very near T$_c$ (ε < 0.01) σ' is closer to log ε than ε$^{-1/2}$.

We stress that the behavior near T_c is little affected by the subtraction procedure to isolate σ' from σ_T. The behavior for very small ε is clearly sensitive to the choice of T_c. (In Samples A, D and E the uncertainty in T_c is ± 0.1 K. This uncertainty is too small to affect the disagreement in **Fig. 4** between the data and the LD curve.)

Discussion of paraconductivity in YBa$_2$Cu$_3$O$_7$ crystals

We will discuss the large ε and small ε regimes separately. The earlier finding that $\sigma' \sim \log \varepsilon$ in the temperature range 30 to 40 K above T_c is an artifact of an arbitrary subtraction procedure, which badly underestimates σ' at temperatures above ~110 K. The new procedure adopted here shows that the $1/\varepsilon$ dependence (Eq. 2) is consistent with the data, or at least cannot be excluded *a priori*. In particular, the $1/T$ behavior at high T in all the 2D models cannot be distinguished from the nominal background, regardless of how high one goes in T. Our procedure also shows that the 3D behavior ($\sigma' \sim \varepsilon^{-1/2}$) can be definitely excluded at large ε. The order-of-magnitude agreement between the value of C determined by this procedure and the theoretical value supports the validity of the new procedure. We note that the good agreement between C and Eq. 2 implies that a substantial fraction of the normal current is undergoing fluctuations into the superconducting state. This fraction is much higher than observed in thin films. To obtain a satisfactory fit to Eq. 7, however, we have had to allow the divergence to occur at T_0, which is a few K lower than T_c (defined as where ρ goes to zero.) This is quite apparent in the plot of $1/\sigma'$ vs T (**Fig. 4.**) The straight line extrapolation of the data intercepts the T axis at T_0, which is clearly a few K below where $1/\sigma'$ first vanishes.

At small ε the paraconductivity is not sensitive to the actual subtraction procedure used, since it becomes larger than σ_N itself. The approach of $1/\sigma'$ to zero is displayed in **Fig. 4**. In Sample 2, $1/\sigma'$ appears to stay on the straight line until ΔT decreases below 0.5 K. Then a fairly sharp transition to zero occurs. This is in contrast to the Lawrence-Doniach theory[19] which predicts a more gentle cross-over to the 3D behavior. From **Fig. 4**, we find that the decrease to zero near T_c is much more abrupt than $\sqrt{\varepsilon}$.

A plausible interpretation of the abrupt jump at T_c is that there might exist a percolative path which first becomes superconducting, thereby shorting out the current. However, we judge this to be highly improbable for the following reasons. If the shorting path is fragile, σ_T should be highly non-Ohmic near T_c. We failed to detect any non-Ohmicity in the T range ($10^{-3} < \varepsilon < 0.2$) up to current densities of 100 A/cm^2. If the shorting path is wide, then the superconducting fluctuations due to this path should dominate the paraconductivity of the lower T_c medium. This is not apparent in **Fig. 4**. Finally, the absence of multiple peaks in the derivative $d\rho_{ab}/dT$ in Samples A, D and E also argues against the existence of large shorting paths.

A more likely interpretation of the striking failure of Eq. 2 is that mean-field (MF) theory is not valid in describing the coupling of fluctuations between adjacent planes very near T_c. The very short ξ_c (1.7 Å $<<$ unit cell dimension along **c**) derived from the attempted fits in **Fig. 4** already indicates a serious flaw in applying the effective-mass approach: Along **c** the GL wave function Ψ varies much too rapidly on the scale of d to justify an effective mass expansion. Hence, the large disagreement near T_c is perhaps to be expected. A closely related issue[22] is the width of the critical region, defined as the range ε_c where the condensation energy per coherence volume is of the order of thermal energy, i.e. $[H_c^2(\varepsilon_c)/8\pi]$. $\xi_{ab}(\varepsilon_c)^2 \xi_c(\varepsilon_c) \sim k_B T_c$. ($H_c$ is the thermodynamic field and ξ_{ab} is the in-plane coherence length.) An interesting possibility is that in the oxide superconductors the critical region is entered *before* the MF crossover to 3D occurs, i.e. the critical region is ~0.5 K in width or $\varepsilon_c \sim 5\times10^{-3}$. Thus, in this respect, the oxide superconductors would be closer to superfluid ^{4}He than to conventional superconductors (where $\varepsilon_c \sim 10^{-10}$.) A very short ξ_c would be consistent with this scenario. In this case σ' is clearly not expected to follow the MF predictions.

No evidence of the 3D ($1/\sqrt{\varepsilon}$) behavior is obtained in the 5 crystals studied. Moreover, the 2D Maki-Thompson contribution[18] which would be ~20 times larger than the AL term [and vary as ($\log \varepsilon$)/ε at large ε] is insignificant.

We gratefully acknowledge useful discussions with Z. Zou, J. Wheatley, T. Hsu, G. Baskaran, P. W. Anderson, and E. Abrahams. This research is supported by the Department of Physics, Princeton University.

References

1. For a survey see Novel Superconductivity edited by Stuart A. Wolf and Vladimir Z. Kresin (Plenum 1987).
2. Proceedings of the Intl. Conf. on High-Temperature Superconductors and Materials and Mechanisms of Superconductivity, Interlaken, 1988, J. Muller & J.L. Olsen, eds., Physica C, to be published.
3. T.K. Worthington et al in Ref. 1, p. 781; T.R. Dinger, T.K. Worthington, W.J. Gallagher, & R.L. Sandstrom, Phys. Rev. Lett. **58**, 2687 (1987).
4. Y. Iye, T. Tamegai, H. Takeya, and H. Takei, Jpn. J. Appl. Phys. **26**. L1057 (1987).
5. H. Noel, P. Gougeon, J.C. Levet, M. Potel, O. Laborde, and P. Monceau, Solid State Commun. **63**, 915 (1987).
6. M. Hikita, Y. Tajima, A. Katsui, Y. Hikada, T. Iwata and S. Tsurumi, Phys. Rev. B **36**, 7199 (1987).
7. S.W. Tozer, A.W. Kleinsasser, T. Penney, D. Kaiser, F. Holtzberg, Phys. Rev. Lett. **59**, 1768 (1987).
8. S.J. Hagen, T.W. Jing, Z.Z. Wang, J. Horvath, & N.P. Ong, Phys. Rev. B **37**, 7928 (1988).
9. Y. Iye et al.in Ref. 2.
10. P.P. Freitas, C.C. Tsuei, & T.S. Plaskett, Phys. Rev. B **36**,833 (1987)
11. M.A. Dubson et al in Ref. 1, p. 981.
12. N. Goldenfeld, P.D. Olmsted, T.A. Friedmann, & D.M. Ginsberg, preprint.
13. B. Oh, K. Char, A.D. Kent, M. Naito, M.R. Beasley, T.H. Geballe, R.H. Hammond, A. Kapitul, & J.M. Graybeal, reported at APS Meeting, New Orleans, March 1988, and preprint.
14. N.P. Ong, Z.Z. Wang, S. Hagen, T.W. Jing, J. Clayhold, & J. Horvath in Ref. 2, to appear.
15. For a review see W.J. Skocpol & M. Tinkham, Rep. Prog. Phys. **38**, 1049 (1975).
16. L.G. Aslamazov and A.I. Larkin, Fiz. Tverd. Tela **10** 1104 (1968) [Sov. Phys.-Solid St. **10** 875 (1968)].
17. Albert Schmid, Zeit. für Physik **215**, 210 (1968).

18. K. Maki, Progr. Theor. Phys. (Kyoto) **39**, 897 (1968); J.E. Crow, R.S. Thompson, M.A. Klenin, and A.K. Bhatnagar, Phys. Rev. Lett. **24**, 371 (1970).
19. W.E. Lawrence & S. Doniach, Proc. 12th Intl. Conf. on Low Temp. Phys., Kyoto, 1970, E. Kanda ed., p. 361.
20. S. Bhattacharya and J. Stokes, private communication.
21. A.T. Fiory, A.F. Hebard, L.F. Schneemeyer, and J.V. Wasczak, to be published; T.T.M. Palstra et al, Phys. Rev. B, to appear.
22. G. Deutscher in Ref. 1, p. 293.

PRECISION MEASUREMENTS ON HIGH T_c SUPERCONDUCTORS

T. SIEGRIST, S. MARTIN, P. MARSH, A. T. FIORY, R. M. FLEMING, L. F. SCHNEEMEYER, S. A. SUNSHINE, AND J. V. WASZCZAK

AT&T Bell Laboratories, 600 Mountain Avenue, Murray Hill, NJ 07974

Abstract

To understand the superconductivity in copper oxide based high T_c superconductors, accurate and precise measurements of their properties are needed. We shall discuss two measurements, (i) an electron density study on $Ba_2YCu_3O_7$ and (ii) the determination of the resistivity tensor in $Bi_2Sr_2CaCu_2O_8$. Both measurements require high precision, and the data have to be corrected for various effects.

The electron density determination in $Ba_2YCu_3O_7$ allows the refinement of anisotropic temperature parameters for all atoms and anharmonic temperature parameters for the metal atoms. In particular, we find a local minimum in the potential well for barium, indicating the possibility of two-site tunneling.

In $Bi_2Sr_2CaCu_2O_8$ a large resistivity anisotropy, of the order of 10^5, is observed for directions parallel and perpendicular to the c-axis. The in-plane resistivity is linear in T, implying two dimensional metallic conduction within the Cu-O planes.

ANHARMONIC AND ANISOTROPIC THERMAL PARAMETERS IN $Ba_2YCu_3O_7$

Superconductivity above 90K has been reported in a number of oxygen-deficient perovskites [1-3]. Several structural studies have identified the superconducting phase as $Ba_2RECu_3O_7$, a perovskite-related structure with 1:1:3 stacking of the cubic aristotype and ordering of RE and Ba cations in the stacking direction [4]. Early single x-ray crystal studies gave a pseudotetragonal structure due to [110] twinning [5-7]. Neutron powder diffraction studies have shown that oxygen ordering is responsible for the orthorhombic symmetry [8-10]. The superiority of the structural refinement in powder studies arises from twinning that cannot be readily resolved in single crystal studies.

Structural studies, using the Rietveld method on neutron powder data, have given precise positional parameters, and have been the method of choice in analyzing oxygen ordering in these materials [8-10]. Also, in some refinements, a split atom position for O 1 was assumed, giving a somewhat better agreement factor. Unfortunately, the Rietveld technique is not suitable for determining thermal motion, even less for anisotropic thermal motion. Accurate Debye-Waller factors of the individual atoms in $Ba_2YCu_3O_7$ are important, since they provide a test for lattice dynamics calculations, and information on the anisotropy of the thermal parameters helps to interpret phonon spectra. In addition, anharmonic behaviour may occur, indicating the presence of local minima in the electrostatic potential, leading to a subtle redistribution of the electron density. In such places, strong polaronic coupling can be expected, that could give rise to an enhanced electron-phonon interaction.

This study presents the results of a multiple wavelength X-ray structural refinement on a single crystal with a small amount of twinning. A model is presented that includes and corrects for the twinning component in the observed data, allowing for refinement in the orthorhombic space group Pmmm. The principal goal of this work is to provide physically meaningful thermal parameters. A key ingredient of the analysis is the simultaneous refinement of the data taken at two x-ray wavelengths. By including the anomalous dispersion at each wavelength, the number of reflections is increased, and we can account for wavelength-dependent features such as extinction and absorption. Therefore, the thermal parameters can be modeled with higher precision. This procedure allows us to refine anisotropic thermal parameters for each atom, including the oxygens. Anharmonic tensors were determined for metal atoms using the Gram-Charlier formalism [11].

Experimental

Single crystals of $Ba_2YCu_3O_7$ were grown as described earlier [12]. A single crystal with {100}, {010} and {001} faces with dimensions of 80✕80✕7.8 μm respectively was used. ω—scans confirmed the presence of a small amount of twinning only (1:10 ratio estimated from the relative intensities of several (h00) and (0k0) reflections), that can be corrected by refining a twinning parameter. All integrated intensities up to $(\sin\theta)/\lambda \leq 1.08\text{Å}^{-1}$ for MoKα radiation and up to $(\sin\theta)/\lambda \leq 0.63\text{Å}^{-1}$ for CuKα radiation, were measured using an Enraf-Nonius CAD-4 diffractometer with graphite-monochromatized radiation. The integrated intensities were corrected for Lorentz, polarization, thermal diffuse scattering (bulk elastic constant of $2\times10^{12} \cdot \text{dyn} \cdot \text{cm}^{-2})$ [13], and absorption effects, as well as for inhomogeneities in the primary beam. The correction for beam inhomogeneity was particularly important, and will be the subject of a separate paper [14]. An extinction correction was applied, using the Becker and Coppens formalism, giving r = 1.27(4) μm as the extinction length.[15]

The lattice constants of $Ba_2YCu_3O_7$ are a = 3.8265(4), b = 3.8833(2) and c = 11.6813(10)Å at 293K, as measured on the CAD-4 diffractometer by measuring absolute θ values of 25 reflections between $19.5 \leq \theta \leq 25.5°$. Two complete data sets for $CuK\alpha$ and $MoK\alpha$ (811 reflections with $(\sin \theta)/\lambda \geq 0.3Å^{-1}$ and $F_m^2 \geq 3\sigma (F_m^2)$) were included in the refinement. The very low angle data were omitted, due to uncertainty in the oxidation state of the individual atoms (difference in the scattering factor > 0.2%). Twinning was included by refining on

$$wR = \left[\Sigma \, w(F_m^2 - xF_{hkl}^2 - (1-x)F_{kh\bar{l}}^2)^2 / \Sigma \, wF_m^4 \right]^{1/2}$$

for x = 0.902(8). The possibility of nonstoichiometry was investigated, but none of the atom occupancies deviated more than two standard deviations from the stoichiometric value. Individual anisotropic temperature parameters for all atoms were refined. Anharmonic temperature parameters were obtained from a Gram-Charlier expansion with

$$t_{GC}(\mathbf{h}) = t_{harm}(\mathbf{h}) \left[1 + \frac{(2\pi i)^3}{3!}\gamma_{ijk}H^{ijk} + \frac{(2\pi i)^4}{4!}\delta_{ijkl}H^{ijkl} \right]$$

where H^{ijk} is a symmetry tensor. For γ_{111}, $H^{111} = h_1^3 a^{*3}$, for γ_{123}, $H^{123} = 6h_1 h_2 h_3 a^* b^* c^*$, etc. As a result, all n-order tensors are expressed in units of $Å^n$. Table I gives the atomic positions obtained, Table II anisotropic thermal parameters and Table III compiles the anharmonic thermal parameters for the barium and copper atoms together with the mean error (in units of σ). Due to the small scattering factor of the oxygen relative to the metal atoms, anharmonic terms for the oxygen atoms were not sufficiently accurate and were not refined.

Results and Discussion

The pseudo-cubic $Ba_2YCu_3O_7$ is a hettotype of the anti-ferroelectric perovskite family, best described by the formula $Ba_2YCu_3O_{9-x}$, for x = 2, where atomic vacancies occur at $(0,0,1/2)$ and $(1/2,0,0)$. Deviations between this structure and the pseudo-cubic aristotypic perovskite are given in Table I. The coordination polyhedron of Cu 1 occurs in point group mmm, precluding tilting of the anion framework or antiparallel displacements of the cation. The average Cu-O distance is 1.90(6)Å, in good agreement with Shannon's radii[6] giving 1.97Å. Foreshortening by 0.042Å occurs for the Cu 1 − O 1 bond. The site symmetry mm of Cu 2 precludes tilting of the anion square pyramid, but severe distortions occur. The Cu 2 − O 2 and Cu 2 − O 3 bond distances compare well with Shannon's radii but the Cu 2 − O 1 distance is inordinately long, 0.353(4)Å longer than that of the other oxygens bonded to Cu 2. The site symmetry of yttrium (mmm) does not allow anti-ferroelectric distortions. The average Y-O distance of 2.40(1)Å compares well with Shannon's radii that give 2.42Å. Barium, with site symmetry mm, shows a displacement of 0.303Å from a plane defined by O 1, but a −0.044Å displacement from an origin defined by the mean location of O 2, O 3, O 4 and □ 1. The Ba - O 1 bond distance of 2.7427(5)Å is extremely short, when compared with Shannon's radii giving 2.92Å. Because of the large size of the Ba atom, it can not occur at the same height as O 1, therefore O 1 is pushed in the −z direction and Ba in the +z direction. Together with Ba,

Table I

Atomic Position Coordinates for $Ba_2YCu_3O_7$ ($\times10^5$)

$a = 3.8265(4)\overset{\circ}{A}$, $b = 3.8833(2)\overset{\circ}{A}$, $c = 11.6813(10)\overset{\circ}{A}$, 295K
Pmmm, R = 0.0167 (811 reflections, 57 variables)

	x	y	z	u ($\overset{\circ}{A}\times10^3$)[a]	$\Delta z(\overset{\circ}{A})$[b]
Ba	50000	50000	18501(7)	93(16)	0.214
Y	50000	50000	50000	73(2)	0
Cu 1	0	0	0	97(12)	0
Cu 2	0	0	35567(7)	71(24)	0.261
O 1	0	0	15902(36)	110(34)	−0.089
O 2	50000	0	37781(38)	94(19)	0.520
O 3	0	50000	37746(34)	82(18)	0.515
O 4	0	50000	0	136(43)	0
□ 1[c]	50000	0	0		0
□ 2	0	0	50000		0

[a] u is the rms radial amplitude of vibration, with standard error given by Bessel's formula.

[b] Δz is the distance between this structure and the equivalent pseudo-cubic aristotypic perovskite.

[c] The symbol □ designates a vacancy which reconciles this structure with the equivalent aristotype.

Table II

Anisotropic temperature parameters for Ba$_2$YCu$_3$O$_7$ (Å $\times 10^3$)[a]

	u_1	u_2	u_3
Ba	109(1)	79(2)	94(1)
Y	79(2)	70(3)	74(2)
Cu 1	112(3)	101(3)	75(3)
Cu 2	66(2)	62(3)	90(2)
O 1	147(8)	113(9)	78(10)
O 2	83(11)	101(10)	109(9)
O 3	94(10)	75(13)	81(11)
O 4	183(13)	105(16)	137(13)

[a] The harmonic temperature factor expression used is
$\exp[-2\pi^2(h^2 a^{*2} u_1^2 + k^2 b^{*2} u_2^2 + l^2 c^{*2} u_3^2)]$

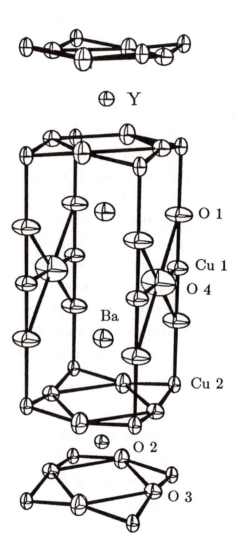

Figure 1   View of a unit cell of $Ba_2YCu_3O_7$. The ellipsoids indicate the thermal motion of each atom.

short Ba - O 1 contact. Since Ba can not vibrate freely towards O 1, the vibration is redirected towards the Cu atoms. The most prominent feature is the large skew function of Ba in the -z direction. Due to the mirror plane at z=0, the related Ba shows the same feature in the +z direction. This accounts for a shortening of the Ba-Ba distance from 4.32Å to 3.42Å, and can be interpreted as a breathing mode along z. Whether this effect represents a static distortion within the crystal, or a dynamic feature involving tunneling or hopping between the two positions cannot be determined, since x-rays average over time. However, arguments for a dynamical feature can be made. The major source of the anharmonic mode is associated with the short Ba - O 1 contact, where foreshortening of 0.017Å occurs at the minimum contact position. This suggests a bimodal electrostatic potential with a small saddle point at z = 0.159. In addition, a Ba atom at $z - \Delta$ will produce a Ba - O 4 contact of 2.56Å, suggesting a highly unstable arrangement, ruling out a static distortion. The Ba — Ba distances due to such a mode are 4.32Å for the +z and —z position, 3.872Å for +z—Δ to —z (+z to —z+Δ) and a very short 3.422Å for +z — Δ to —z + Δ. The very short distance for the latter makes this a highly unstable arrangement, and is expected to be sparsely populated. From the observed density, we estimate that 2 to 4% of the barium atoms occupy the local potential well. Cu 1 shows some anharmonic motion in the basal plane. Since Cu 1 is restricted in its motion towards O 1, it deforms in a direction perpendicular to the bond. Also, Cu 2 displays a slightly skewed electron density along the c-axis, since the motion in the xy plane is restricted by the O 2 and O 3 oxygen atoms.

Table III

Anharmonic Thermal Parameters for $Ba_2YCu_3O_7(\mathring{A}^n \times 10^n)^a$

	Ba	$\dfrac{u}{\sigma(u)}$	Cu 1	$\dfrac{u}{\sigma(u)}$	Cu 2	$\dfrac{u}{\sigma(u)}$
γ_{333}	−0.028	1.0			0.316	7.5
γ_{113}	0.241	14.8			0.010	0.3
γ_{223}	−0.012	0.9			0.035	1.2
δ_{1111}	1.009	2.3	3.409	6.4	−1.099	5.5
δ_{2222}	−0.877	8.3	7.160	11.0	0.024	0.1
δ_{3333}	0.099	1.1	1.133	3.6	−1.358	6.1
δ_{1122}	0.046	1.1	1.914	9.8	0.099	1.3
δ_{1133}	−0.007	0.1	0.139	0.8	0.017	0.2
δ_{2233}	−0.193	5.7	1.076	6.0	0.250	2.6

[a] Anharmonic thermal parameters involve the Gram-Charlier formalism.

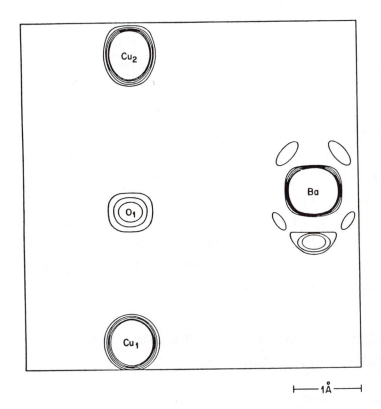

Figure 2   An electron density map of $Ba_2YCu_3O_7$ for a plane
involving Cu 1, Cu 2 and Ba. The contours are from
10 to 50 e$\overset{\circ}{A}^{-3}$.

Unfortunately, the weak scattering power of the oxygen atoms does not allow any meaningful refinement of anharmonic tensors for these atoms. Anharmonic motion is particularly expected for O 4, since this oxygen atom can easily be removed by heating.

In summary, we have obtained anisotropic thermal parameters for $Ba_2YCu_3O_7$ from a dual wavelength x-ray crystal structure determination using a single crystal with a small amount of twinning. In addition, anharmonic thermal motion was found for Ba, Cu 1 and Cu 2, whereas Y does not show any anharmonicity. The anisotropic and anharmonic thermal parameters reflect the particular coordination polyhedron, as well as the bonding distances involved. The precision of these results is due to the simultaneous refinement of data collected at two different wavelengths.

TEMPERATURE DEPENDENCE OF THE RESISTIVITY TENSOR IN $Bi_2Sr_2CaCu_2O_8$ CRYSTALS

Among the fascinating aspects of the known high-T_c superconductors [17-19] are the anisotropic layered structure, the nearly two-dimensional (2D) confinement of the carriers to Cu-O planes, and their implications for the mechanisms of electrical transport and superconductivity. The recently discovered $Bi_2Sr_2CaCu_2O_8$ class of superconductors [20-24] is striking by a 12 Å spacing between the Cu-O planes, which is larger than in the $(La,Sr)_2CuO_4$ and $YBa_2Cu_3O_x$ superconductors, and in itself suggests an enhanced two-dimensional character of electronic properties. Moreover, $Bi_2Sr_2CaCu_2O_8$ is found to be thermally stable with respect to oxygen stoichiometry, and can be grown as single crystals in the form of thin sheets oriented with the c axis normal. By comparison, the oxygen intercalation in superconducting $Ba_2YCu_3O_7$ can be inhomogeneous, and the larger orthorhombicity leads invariably to microscopic twinning.

In this work we report measurements of the resistivity tensor components ρ_a, ρ_b and ρ_c in orthorhombic crystals of nominal composition $Bi_2Sr_{2.2}Ca_{0.8}Cu_2O_8$ (estimated from Rutherford backscattering analyses), which show more pronounced anisotropy than the other oxide superconductors: [25-29] $\rho_c/\rho_a = 1.5 \cdot 10^5$ at T_c, decreasing to about $5 \cdot 10^4$ at 600 K. The a b plane resistivity is also anisotropic, with a ratio of $\rho_a/\rho_b \approx 1.7$. These new results can be explained by 2D transport within the Cu-O planes which are nearly insulated from one another, giving a resistivity linear in T.[30] The high value of ρ_c is consistent with the larger separation between Cu-O planes in $Bi_2Sr_2CaCu_2O_8$ The non-metallic, approximately T^{-1}, term in ρ_c may be indicative of interplanar quasiparticle tunneling, as proposed by Anderson et al.[31] However, we present an alternative argument, that the observed conductivity along the c axis may not be intrinsic, if there are topological defects acting as local short circuits distributed throughout the crystal.

Single crystals used in our study were grown in alkali-chloride fluxes in the form of thin platelets.[32] Results described here were obtained on a crystal of dimensions $L_a = 0.6$ mm, $L_b = 1.1$ mm, and $L_c = 1\,\mu$m. The samples were oriented by x-ray diffraction, which shows the characteristic superlattice splittings only along [0k0] and a peak at (300) but not at (030). As shown schematically as insets in Fig. 3, Au wires (25 μm dia.) were attached at 6 points on the crystal using a Au-paste and Ag-paint

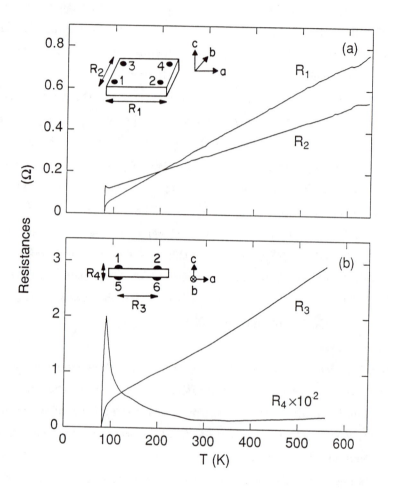

Figure 3   (a)   Temperature dependence of the resistances measured along the a- and b-axes of a $Bi_2Sr_2CaCu_2O_8$ single crystal. The inset is a schematic of the contact configuration. Onset of superconductivity occurs at T = 90 K and zero resistance is reached at $T_c = 81$ K.
(b)   Temperature dependence of the resistances measured along the a- and c-axes.
The lines are drawn through closely-spaced data points.

mixture which was cured by heating to $\sim300°C$ in dry O_2. Contacts were monitored to verify stable ohmic $(0.5-5m\Omega cm^2)$ behavior with temperature cycling. Measurements were performed in a single oven/cryostat using O_2 above 300 K and He below. Resistances were measured using a standard ac phase-sensitive technique with $350\,\mu A$ excitation current and $f = 11.3$ Hz.

Figure 3a shows data obtained from measurements in the a b plane. Referring to the contact configuration, resistance R_1 was measured by passing current through contacts 1 and 2 (a-axis) and measuring the potential difference between 3 and 4. R_2 was obtained by a 90° rotation of the configuration (b-axis). We found zero resistance at $T_c = 81$ K, an onset of the superconducting transition at about $T = 90$ K, and the absence of higher-T_c phases in selected samples. Both R_1 and R_2 show less than 10% deviation from a linear temperature dependence up to 600 K.

Figure 3 b shows the results of measurements performed along the a- and c-directions. The resistance R_3 along the a-direction increases monotonically with temperature, but shows a clear deviation from linearity. R_4 rises sharply at T_c, and then decreases with increasing temperature. A strong temperature dependence for R_4 is seen only between 100 K and 250 K. We emphasize, however, that the measured resistances are influenced by the geometry of the contacts, and do not necessarily reflect the true behavior of the resistivities.

To make the transformation from the R_i measurements to the ρ_i components of the resistivity tensor we followed the basic method of Montgomery [33] and Logan et al., [34] which maps the anisotropic crystal of physical dimensions L_i into an equivalent isotropic block of scaled dimensions l_i. Owing to the small L_c

dimension of the crystals and the placement of inevitably finite-size contacts, the Montgomery analysis should not be used in its original form, as it assumes that contacts are located at the corners of a crystal face. We therefore generalized the method of images solution presented by Logan et al., upon which Montgomery's analysis is based, to simulate the actual points of electrical contact. The lattice sums[34] were evaluated by Ewald's method.[35] We determined calibration curves giving the dependence of the resistance ratios R_i/R_j, corresponding to a specific configuration of four contacts in crystal plane 'k' (cyclic order), as functions of effective length ratios l_i/l_j. These functions have properties similar to those of Montgomery and Logan et al., except that they lack inversion symmetry. Our procedure was also used to test effects of contact misalignment and uncertainty in the contact positions, owing to the finite contact size. Since we employed six contacts, the resistance-ratio data was partially redundant, and gave us a consistency check. It confirmed the accuracy of our contact coordinates on the crystal. We then derived l_a/l_b and l_c/l_a from the two sets of data R_1/R_2 and R_4/R_3. The final step was to calculate resistivity ratios according to the scaling formula: $\rho_i/\rho_j = (l_i/l_j)^2 \cdot (L_j/L_i)^2$.

The results of the deconvolution of the data are shown in Fig. 4, which gives the temperature dependence above T_c for the three independent resistivity ratios ρ_a/ρ_b, ρ_c/ρ_a and ρ_c/ρ_b. The results indicate the presence of a temperature–dependent ρ_a/ρ_b resistivity anisotropy in the a b plane, which decreases significantly for $T \lesssim 300\,K$. Measurements in several samples yielded values for the a b anisotropy ranging from 1.5 up to about 2, which is in reasonable agreement with results obtained from epitaxial films of

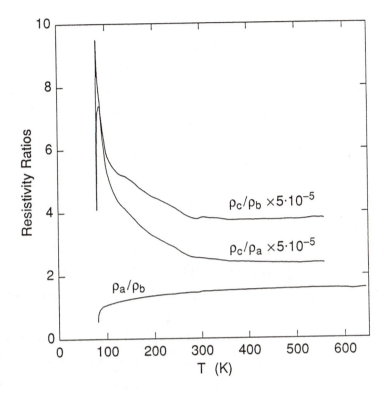

Figure 4   The in-plane (ρ_a/ρ_b) and out-of-plane $(\rho_c/\rho_a$ and $\rho_c/\rho_b)$ resistivity ratios calculated from the measured data of Fig. 1. The anisotropies are ≈ 1.7 and $\approx 10^5$, respectively.

$Bi_2Sr_2CaCu_2O_8$.[36] The ratios ρ_c/ρ_a and ρ_c/ρ_b are found to decrease monotonically with increasing temperature, as shown in Fig. 4. We note a strong temperature dependence from T_c to about 275 K and saturation behavior above 275 K.

When one of the l_i is smaller than the other two, i.e. $l_k <$ $(l_i l_j)^{1/2}$, one can analyze the R_i/R_j ratio corresponding to $l_i < l_j$. In this case one obtains the geometric average $\rho_{ij} = (\rho_i \rho_j)^{1/2} =$ $R_i L_k \cdot f(R_i/R_j, l_k)$ which is a generalization of the van der Pauw analysis.[33] For the data of Fig. 1, the electrically thinnest dimension turns out to be L_c. Figure 5a gives the temperature dependences of ρ_{ab} and ρ_c/ρ_{ab} derived in this manner. Figure 5b shows resistivity tensor components which were calculated from ρ_{ab} and the resistivity ratios (Fig. 4). The uncertainty in the resistivity values scales with that of the crystal thickness (factor ~2).

The averaged a b plane resistivity ρ_{ab} increases proportionally to the temperature with a slope of $\alpha_{ab} = 0.46 \mu\Omega cm K^{-1}$. At ambient temperature we have $\rho_{ab} = 140 \mu\Omega cm$, in good agreement with previous reports.[24] We find that ρ_a and ρ_b also increase linearly over the complete temperature range. The extrapolation to T = 0 K yields residual resistivities of less than 10 $\mu\Omega cm$. The resistivity along the c-axis is found to be unusually large, with a linear temperature dependence commencing above 275 K. Below 275 K an increase of ρ_c above the linearly extrapolated values can be clearly seen. As in $Ba_2YCu_3O_7$ crystals, the absence of saturation·behavior leads to a weak electron-phonon coupling constant $\lambda \lesssim 0.2$; from α_{ab} we estimate a transport plasma energy of $\Omega_{ab} \lesssim 1.3eV$, and a London penetration depth of $\lambda_{ab} \lesssim 1400 Å$.[37]

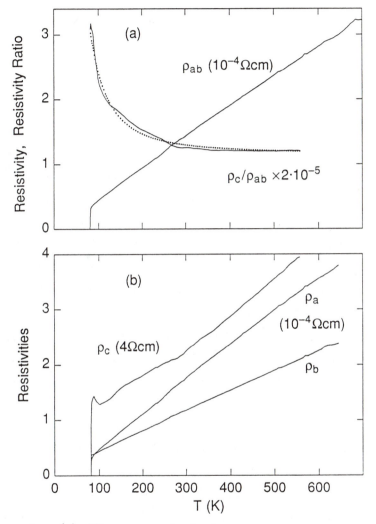

Figure 5   (a) The averaged a b-plane resistivity ρ_{ab} and the ratio ρ_c/ρ_{ab} calculated from the measured data as described in the text. The dotted line is the result of a 2-parameter fit $\rho_c/\rho_{ab} = a + b\,T^{-2}$.
(b) The three components of the resistivity tensor as a function of temperature.

In contrast with data from $Ba_2YCu_3O_7$ single crystals, we find that $Bi_2Sr_2CaCu_2O_8$ samples exhibit a resistive anisotropy perpendicular to the Cu-O planes which is three orders of magnitude larger. If we consider each Cu-O plane to be an electrically-isolated metallic sheet, with 4 sheets per lattice constant c, then the sheet resistance is $R_{ab} = \rho_{ab}(c/4)^{-1} \approx 300\,\Omega/\square$ just above T_c. This is low enough to allow for metallic behavior and the occurrence of two-dimensional superconductivity, apart from phase fluctuations.[38] The calculation for $Ba_2YCu_3O_7$ yields $4000\Omega/\square$, which is too large to agree with our first assumption of a good 2D superconductor.[38] The magnitude of the inter-planar coupling can now be estimated by relating the difference in the spacing between planes for $Ba_2YCu_3O_7$ (d = 8.3 Å) and $Bi_2Sr_2CaCu_2O_8$ (d = 12 Å) with the anisotropy ratio being smaller in $Ba_2YCu_3O_7$ by 10^3. If the anisotropy ratio is proportional to the transmission coefficient t : $\rho_c/\rho_{ab} \propto t \propto \exp[-\kappa d]$, where κ is the decay parameter for tunneling through Cu-0 planes involving similar potential barriers ($\sim$2eV), then we arrive at $\kappa = 1.6\text{Å}^{-1}$.

Recently, Hagen et al. [36] showed that ρ_c in $Ba_2YCu_3O_7$ contains a T^{-1} term, which Anderson et al. [31] have interpreted as evidence for a quasiparticle tunneling mechanism of electrical transport between planes. The authors proposed that the measured ρ_c contains a linear-T term from ρ_{ab}, owing to contact misalignment. The T^{-1} dependence of the intrinsic ρ_c and the T dependence of ρ_{ab} combine, such that the ratio ρ_c/ρ_{ab} has a T^{-2} term. Recognizing the possibility of mixing, we try to reveal this divergence in a transparent manner by plotting in Fig. 5a the ratio ρ_c/ρ_{ab}, a quantity which is similar to the resistivity ratios presented in Fig. 4. The dotted curve in Fig. 5a shows the result of a linear fit $\rho_c/\rho_{ab} = a + bT^{-2}$, but which is plotted on the linear-T scale, to give intrinsic $\rho_c = 300\Omega\text{cmK}\cdot T^{-1}$, and a mixing fraction $a^{-1} = 2\cdot10^{-5}$. Assuming a power-law behavior, we obtain a T^{-1} dependence to within 10% systematic error in the -1 exponent.

In this present case our numerical analysis could not eliminate the saturation behavior of ρ_c/ρ_{ab} curve by repositioning the contact points. This finding, and the difficulty in explaining large observed resistivities together with metallic temperature dependences, suggest that in general an alternative explanation should be considered. We postulate that current transport along the c-axis could be dominated by defects acting as links between otherwise electrically insulated a b planes. These defects could be comprised of missing or extra sublayers, stacking faults or dislocations, which have been recently observed by transmission electron microscopy studies in $Bi_2Sr_2CaCu_2O_8$ [39] and in $Ba_2YCu_3O_7$,[40] indicating that defect structures may be relatively common.

A simple model starting with unity anisotropy, and attributing the large ρ_c for $T > T_c$ and $\rho_c = 0$ for $T < T_c$ to gross exfoliation of the material, cannot correctly account for a temperature dependence in the resistivity ratios. Instead, we assume there is a random network of marginally metallic paths, caused by an unspecified defect topology along c, which we represent by a regular array of links, spaced ξ apart in the a b directions, and spanning the mean distance d_q between planes. Representing the link resistance by R_d and the a b-plane resistance between links as R_{ab}, we can see that the apparent resistivity along the c direction becomes $\rho_c = (R_d + R_{ab})(\xi/2)^2 d_{\P}^{-1}$. We take $R_{ab} = \rho_{ab} d_{\P}^{-1}$, and obtain $\rho_c/\rho_{ab} = (1 + R_d/R_{ab}) \cdot (\xi/2d_{\P})^2$. From the data of Fig. 3a we can conclude that $R_{ab} \gg R_d$ for $T \gtrsim 275\,K$, and the saturation yields $\xi \approx 500\,d_{\P} = 0.6\mu m$. The apparent divergence at low temperatures could be attributed to a negative temperature coefficient of R_d, arising simply from a distribution of activated conducting paths.

In conclusion, we have measured the resistivity tensor of $Bi_2Sr_2CaCu_2O_8$ single crystals as a function of temperature. We find an a b-plane anisotropy $\rho_a > \rho_b$, which shows that the highest conductivity is along the superlattice. In the c direction the large anisotropy factor of about 10^5 is interpreted in terms of a highly two-dimensional character of the electronic structure. Although the T^{-1} divergence of ρ_c in our data at low temperatures could be accounted for by quasiparticle tunneling between planes, we caution that the high resistivity also implies high sensitivity to bridging defects, which can short circuit the Cu-O planes. Various transport parameters are derived from the data, most importantly an inter-planar tunneling constant, a b-plane transport plasma energy, London penetration depth, and the electron-phonon coupling constant.

We would like to acknowledge many useful discussions with our colleagues, D.W. Murphy, B. Batlogg, R.J. Cava, R.B van Dover, A.F. Hebard, R. Hull, P.B. Littlewood, P. Marsh, A.J. Millis, S. Nakahara, T.T. Palstra and A.P. Ramirez. One of us (S.M.) acknowledges partial support by the Alexander von Humboldt Foundation, West Germany.

REFERENCES

1. R. J. Cava, B. Batlogg, R. B. van Dover, D. W. Murphy, S. A. Sunshine, T. Siegrist, J. P. Remeika, E. A. Reitman, S. Zahurak, and G. P. Espinosa, Phys. Rev. Lett. 58, 1676 (1987).

2. J. M. Tarascon, W. R. McKinnon, L. H. Greene, G. W. Hull and E. M. Vogel, Phys. Rev. B36, 226 (1987).

3. L. F. Schneemeyer, J. V. Waszczak, S. M. Zahurak, R. B. van Dover, and T. Siegrist, Mater. Res. Bull. 22, 1467 (1987).

4. Y. LePage, T. Siegrist, S. A. Sunshine, L. F. Schneemeyer, D. W. Murphy, S. M. Zahurak, J. V. Waszczak, W. R. McKinnon, J. M. Tarascon, G. W. Hull, and L. H. Greene, Phys. Rev. B36, 3617 (1987).

5. T. Siegrist, S. Sunshine, D. W. Murphy, R. J. Cava, and S. M. Zahurak, Phys. Rev. B35, 7137 (1987).

6. Y. LePage, W. R. McKinnon, J. M. Tarascon, L. H. Greene, G. W. Hull, and D. M. Hwang, Phys. Rev. B35, 7245 (1987).

7. R. M. Hazen, L. W. Finger, R. J. Angel, C. T. Prewitt, N. L. Ross, H. K. Mao, C. G. Hadidiacos, P. H. Hor, R. L. Meng, and C. W. Chu, Phys. Rev. B35, 7238 (1987); B36, 3966(E) (1987).

8. J. E. Greedan, A. O'Reilly, and C. V. Stager, Phys. Rev. B35, 8770 (1987).

9. F. Beech, S. Miraglia, A. Santoro, and R. S. Roth, Phys. Rev. B35, 8778 (1987).

10. J. J. Capponi, C. Chaillout, A. W. Hewat, P. Lejay, M. Marezio, N. Nguyen, B. Raveau, J. L. Soubeyroux, J. L. Tholence, and R. Tournier, Europhys. Lett. 3, 1301 (1987).

11. U. H. Zucker and H. Schulz, Acta Crystallogr. A38, 563 (1982); International Tables for Crystallography, Vol. IV, 311 (Kynoch Press, Birmingham, 1974).

12. L. F. Schneemeyer, J. V. Waszczak, T. Siegrist, R. B. van Dover, L. W. Rupp, B. Batlogg, R. J. Cava, and D. W. Murphy, Nature 328, 601 (1987).

13. D. J. Bishop, A. P. Ramirez, P. L. Gammed, B. Batlogg, E. A. Rietman, R. J. Cava, and A. J. Millis, Phys. Rev. B36, 2408 (1987).

14. P. Marsh

15. J. P. Becker and P. Coppens, Acta Crystallogr. A30, 129 (1974).

16. R. D. Shannon, Acta Crystallogr. A32, 751 (1976).

17. J. G. Bednorz and K. A. Müller, Z. Phys. B 6, 189 (1986).

18. R. J. Cava, R. B. van Dover, B. Batlogg, and E. A. Rietman, Phys. Rev. Lett. 58, 408 (1987).

19. M. K. Wu, J. R. Ashburn, C. J. Torng, P. H. Hor, R. L. Meng, L. Gao, Z. J. Huang, Y. Q. Wang, and C. W. Chu, Phys. Rev. Lett. 58, 908 (1987).

20. C. Michel, M. Hervieu, M. M. Borel, A. Grandin, F. Deslandes, J. Provost, and B. Raveau, Z. Phys. B 68, 421 (1987).

21. H. Maeda, Y. Tanaka, M. Fukutomi, and T. Asano, submitted for publication.

22. S. A. Sunshine, T. Siegrist, L. F. Schneemeyer, D. W. Murphy, R. J. Cava, B. Batlogg, R. B. van Dover, R. M. Fleming, S. H. Glarum, S. Nakahara, R. Farrow, J. J. Krajewski, S. M. Zahurak, J. V. Waszczak, J. H. Marshall, P. Marsh, L. W. Rupp, Jr., and W. F. Peck, Phys. Rev. B, submitted (1988).

23. C. W. Chu, J. Bechtold, L. Gao, P. H. Hor, Z. J. Huang, R. L. Meng, Y. Y. Sun, Y. Q. Wang, and Y. Y. Xue, Phys. Rev. Lett. 60, 941 (1988).

24. H. Takagi, H. Eisaki, S. Uchida, A. Maeda, S. Tajima, K. Uchinokura, and S. Tanaka, submitted for publication.

25. S. W. Tozer, A. W. Kleinsasser, T. Penney, D. Kaiser, and F. Holtzberg, Phys. Rev. Lett. 59, 1768 (1987).

26. S. J. Hagen, T. W. Jing, Z. Z. Wang, J. Horvath, and N. P. Ong, Phys. Rev. B, in print (1988).

27. S. W. Cheong, Z. Fisk, R. S. Kwok, J. P. Remeika, J. D. Thompson, and G. Grüner, submitted for publication.

28. K. Murata et al., Jpn. J. Appl. Phys. 26, L1941 (1987); the authors point out striking similarities to earlier work on resistance anisotropy in layered low-T_c superconductors.

29. B. I. Verkin et al., Fizika Nizkikh Temperatur 14, 218 (1988) (Sov. J. Low Temp. Phys.); the authors' work indicates a large a b c anisotropy in untwinned $YBa_{2-x}Sr_xCu_3O_{7-y}$ crystals.

30. R. Micnas, J. Ranninger, and S. Robaszkiewicz, Phys. Rev. B36, 4051 (1987).

31. P. W. Anderson, G. Baskaran, Z. Zou, and T. Hsu, Phys. Rev. Lett. 58, 2790 (1987); P.W. Anderson and Z. Zou, Phys. Rev. Lett. 60, 132 (1988).

32. L. F. Schneemeyer, R. B. van Dover, S. H. Glarum, S. A. Sunshine, R. M. Fleming, B. Batlogg, T. Siegrist, J. M. Marshall, J. V. Waszczak, and L. W. Rupp, Nature 332, 422 (1988).

33. H. C. Montgomery, J. Appl. Phys. 42, 2971 (1971).

34. B. F. Logan, S. O. Rice, and R. F. Wick, J. Appl. Phys. 42, 2975 (1971).

35. J. M. Ziman, Principles of the Theory of Solids, (Cambridge University Press, London, 1972), p.37.

36. C. E. Rice, A. F. J. Levi, A. F. Hebard, private communications.

37. M. Gurvitch and A. T. Fiory, Phys. Rev. Lett. 59, 1337 (1987).

38. A. F. Hebard and M. A. Paalanen, Phys. Rev. B30, 4063 (1984).

39. R. M. Hazen et al., Phys. Rev. Lett. 60, 1174 (1988); D. R. Veblen et al., Nature 332, 334 (1988).

40. H. W. Zandbergen, R. Gronsky, K. Wang, and G. Thomas, Nature 331, 596 (1988).

LOW-TEMPERATURE DENSIFICATION OF HIGH-T_c SUPERCONDUCTORS[1]

D. W. CAPONE II[2]
Argonne National Laboratory, Argonne, IL 60439, USA

Abstract. It is believed that the weak-link behavior in $YBa_2Cu_3O_7$, "123" bulk materials results from the presence of non-superconducting second phases coating the grain boundaries of sintered "123" compacts. These second phases result from the $BaCuO_2$ - CuO eutectic, which is a liquid at sintering temperatures above 870°C. Sintering below this temperature results in low densities (ca. 70% of the theoretical density). Sintering above 870°C produces dense samples via liquid-phase sintering, resulting in the grain boundary phases mentioned above. In this talk we report the results of a series of low-temperature densification experiments designed to produce 100% dense ceramic samples at temperatures below 870°C. Room-temperature swaging, using standard powder metallurgy techniques, has been used to form fibers having "green" (unsintered) densities greater than 90% of the theoretical density of the "123". A warm extrusion technique ($T \geq 300$°C) has also been used to produce 100% dense "123" compacts. In both cases, after fabrication, the oxygen content of the "123" materials is near 7.0. However, x-ray diffraction experiments show that the cation lattice is disordered, and the materials are not superconducting. These materials can be ordered using suitable post-annealing techniques. The effect of time and temperature on the degree of ordering and the superconductivity will be presented.

---

1) Work supported by the U.S. Department of Energy, BES-Materials Sciences, under Contract #W-31-109-ENG-39
2) Present address: Supercon, Inc., 830 Boston Turnpike Rd., Shrewsbury MA 01545, USA.

ANOMALOUS MAGNETIC, PHOTOINDUCED, AND STRUCTURAL PHENOMENA IN HIGH-T$_c$ MATERIALS

A. J. EPSTEIN,[a] X. D. CHEN,[a] S. CHITTIPEDDI,[a] E. EHRENFREUND†,[a] J. M. GINDER,[a] Y. LU,[a] R. P. McCALL,[a] B. R. PATTON,[a] M. G. ROE,[a] F. ZUO,[a] W. FARNETH,[b] B. S. McLEAN,[b] S. I. SHAH,[b] J. R. GAINES[c] and Y. SONG[c]

a) Department of Physics, The Ohio State University, Columbus, Ohio 43210, U.S.A.
b) Central Research and Development Department, Experimental Station, E. I. du Pont de Nemours, Inc., Wilmington, Delaware 19808, U.S.A.
c) Department of Physics, University of Hawaii, Honolulu, Hawaii 96844, U.S.A.

Abstract The implications of photoinduced absorption and bleaching, of the effect of high magnetic fields, and of chemical substitutions in La$_{2-x}$Sr$_x$CuO$_{4-\delta}$ and R$_1$Ba$_2$Cu$_3$O$_{6+x}$ (R = Y,. rare earth) are discussed.

INTRODUCTION

The discovery of high-temperature superconductivity in the La-Ba-Cu-O system[1] has stimulated extensive study of these and related materials. Numerous models have now been proposed invoking the role of electron-phonon, electron-electron, and excitonic mechanisms in the origin of the unusually high temperature superconductivity.[2-9] Of the several families of high-temperature superconductors, two of them, La$_{2-x}$Sr$_x$CuO$_{4-\delta}$ and R$_1$Ba$_2$Cu$_3$O$_{6+x}$ (where R is Y, or a rare earth), can be varied from a magnetic insulating state to a metallic superconducting state by increase in the Sr or O content, respectively. We briefly discuss

---

†Permanent address: Department of Physics, Technion, Haifa, Israel

here the results of studies using three techniques to probe the nature of the electronic states in the insulating form: photoinduced absorption spectroscopy, high magnetic field studies, and chemical substitution. These results give insight into the nature of the states in the insulating form, and their implications for the formation of superconducting states.

PHOTOINDUCED ABSORPTION AND BLEACHING

We have performed an experimental study of the photoinduced optical phenomena in La_2CuO_4 for probe-photon energies between 0.4 and 2.4 eV, and a pump-laser-photon energy of 2.54 eV. Several features were observed for the first time:[10] luminescence at ∼ 2 eV, confirming the presence of an energy gap at ∼ 2 eV; *two* photoinduced absorption peaks, at 0.5 and at 1.4 eV; and a crossover from photoinduced absorption to photoinduced bleaching near 2 eV. The observation of a 0.5 eV photoinduced absorption confirms previously reported results.[11] The intensites of the absorption and bleaching bands all decrease similarly with increasing temperature. The intensities of the photoinduced features are roughly proportional to the square root of the incident laser power, indicating a bimolecular type decay, though the chopping frequency behavior of the two photoinduced absorption peaks differs somewhat.

Several of the models that have been proposed for elecronic state formation in the layered superconductors may be relevant for these excitations, including solitons in a resonating valence bond lattice,[7] spin bags within an antiferromagnetic ground state,[9] and bipolarons within a two-dimensional charge density wave ground state.[12] It is intriguing to note that the photoexcited defect absorptions we observe occur   at energies essentially identical to those of doping-induced defects.[13-15] Thin films of $YBa_2Cu_3O_{6+x}$ also exhibit photoinduced effects, with bleaching generally observed for energies > 2 eV, although the details of the photoinduced spectrum are quite sensitive to sample preparation and handling history.[16]

EFFECT OF HIGH MAGNETIC FIELDS

Application of high magnetic fields to antiferromagnetic samples of La_2CuO_4 induces an anomalous magnetic behavior not present in samples that have no magnetic ordering. For temperatures $T > T_N$, where T_N is the Neel temperature, the magnetization M exhibits a sublinear dependence on magnetic field H. For $T < T_N$ there is an anomalous increase in M at a T-dependent critical field, approximately 60 kG.[17] This behavior is in accord with the existence of ferromagnetic domains above T_N, which become antiferromagnetically coupled into the surrounding lattice below T_N, and subsequently undergo a transition back to the ferromagnetic state under application of a sufficiently large applied field. These results suggest the importance of local magnetic order around defects within the otherwise antiferromagnetically ordered CuO system.

CHEMICAL SUBSTITUTION

It is well known that substitution of different rare earths in the $R_1Ba_2Cu_3O_{6+x}$ system produces nearly identical systems, with the exceptions of Tb, Ce, and Pr. All three of these latter rare earths have a tendency toward valence instability, which may relate to the difficulties in making superconductors from these compounds. In order to test the effects of local valency, we have made a comparative study of $Nd_1Ba_2Cu_3O_{6+x}$ and $Pr_1Ba_2Cu_3O_{6+x}$. These systems differ by only one atomic number, though the neodymium compound is superconducting, while the praseodymium compound is semiconducting. Detailed resistive and magnetic studies of these systems suggest that the differences relate to charge-carrier density and the possible role of valence fluctuations.[18]

CONCLUSION

In summary, extensive studies of localized states in the superconducting rare earth families of materials show the important role of local structure and local defect structure in the formation of the metallic and superconducting states.

ACKNOWLEDGEMENT

This work is supported in part by the Defense Advanced Research Projects Agency through a grant monitored by the U. S. Office of Naval Research and the Department of Energy Grant Number DE-FG02-86ER45271.A000.

REFERENCES

1. G. Bednorz and K. A. Müller, Z. Phys. B, 64, 189 (1986).
2. P. W. Anderson, Science, 235, 1196 (1987).
3. P. W. Anderson, G. Baskaran, Z. Zou, and T. Hsu, Phys. Rev. Lett., 58, 2790 (1987).
4. V. J. Emery, Phys. Rev. Lett., 58, 2794 (1987).
5. P. A. Lee and M. Read, Phys. Rev. Lett., 58, 2691 (1987).

6. J. E. Hirsch, Phys. Rev. Lett., 59, 228 (1987).
7. S. Kivelson, D. S. Rokhsar, and J. P. Sethna, Phys. Rev. B, 35, 8865 (1987).
8. D. J. Thouless, Phys. Rev. B, 36, 7187 (1987).
9. J. R. Schrieffer, X. G. Wen, and S. C. Zhang, Phys. Rev. Lett., 60, 944 (1988).
10. J. M. Ginder, M. G. Roe, Y. Song, R. P. McCall, J. R. Gaines, E. Ehrenfreund, and A. J. Epstein, Phys. Rev. B, 37, 7506 (1988).
11. Y. H. Kim, A. J. Heeger, L. Acedo, G. Stucky, and F. Wudl, Phys. Rev. B, 36, 7252 (1987).
12. P. Prelovsek, T. M. Rice, and F. C. Zhang, J. Phys. C, 20, L229 (1987).
13. S. L. Herr, K. Kamaras, C. D. Porter, M. G. Doss, D. B. Tanner, D. A. Bonn, J. E. Greedan, C. V. Stager, and T. Timusk, Phys. Rev. B, 36, 733 (1987).
14. S. Etemad, D. E. Aspnes, M. K. Kelly, R. Thompson, J. -M. Tarascon, and G. W. Hull, Phys. Rev. B, 37, 3396 (1988).
15. M. Suzuki, Bull. Am. Phys. Soc., 33, 213 (1988).
16. J. M. Ginder et al., to be published.
17. F. Zuo, X. D. Chen, J. R. Gaines, and A. J. Epstein, Phys. Rev. B, 38, 901 (1988).
18. S. Chittipeddi, Y. Song, D. L. Cox, J. R. Gaines, J. P. Golben, and A. J. Epstein, Phys. Rev. B, 37, 7454 (1988).

SUPERCONDUCTIVITY AND MAGNETISM OF La-Cu-O COMPOUNDS WITH K_2NiF_4 STRUCTURE

K. SEKIZAWA, Y. TAKANO, S. TASAKI, N. YAMADA* AND
T. OHOYAMA*
College of Science & Technology, Nihon University,
Chiyoda-ku, Tokyo 101, JAPAN
*The University of Electro-Communications, Chofu-shi,
Tokyo 182, JAPAN

Abstract   A superconducting transition with the onset
temperature of about 40 K was observed in undoped
La_2CuO_z with the distorted K_2NiF_4 type structure for
certain sample preparation conditions. This superconductivity occurs in samples made from insufficiently
mixed powders, and also in oxygen-treated samples. The
essential factor for the occurrence of superconductivity
in undoped La_2CuO_z is not the homogeneous nonstoichiometry, but the existence of inhomogeneity in the sample.
A similar situation is realized in oxygen-treated
samples by the introduction of excess oxygen into the
lattice.

INTRODUCTION

Superconductivity may occur in the undoped La-Cu-O system when
a proper amount of carriers is introduced into the system.
As the number of carriers may change with n in the series of
perovskite-related compounds $LaO(LaCuO_3)_n$, we can expect
superconductivity even in the La-Cu-O compounds undoped with
alkaline earth ions. We tried to synthesize the compounds
with $n = 1,2$ and 3 in this series, and investigated their
crystallographic and superconducting properties. In the
course of our study,[1,2] however, we have observed superconductivity even in La_2CuO_z prepared under certain conditions.

TABLE I.   Lattice parameters of La-Cu-O compounds.

samples	degree of homo-geneity	O_2 treat-ment	a(A)	b(A)	c(A)
#1 La_2CuO_4	A	---	5.357	5.397	13.14
#2 La_2CuO_z	B	---			
#3 $La_2Cu_{1.04}O_z$	A	---	5.356	5.405	13.15
#4 $La_2Cu_{0.96}O_z$	A	---	5.357	5.408	13.15
#5 La_2CuO_z	A	o	5.358	5.394	13.17
#6 La_2CuO_z	B	o			
#7 $La_2Cu_{1.04}O_z$	A	o	5.355	5.394	13.16
#8 $La_2Cu_{0.96}O_z$	A	o	5.356	5.392	13.15

In the following, we report mainly about this phenomenon.

SAMPLES

Samples with nominal compositions $La_2Cu_{1+x}O_z$ ($x = 0, \pm 0.04$),
were prepared by the solid-state reaction. Appropriate
amounts of $4N-La_2O_3$ and $3N-CuO$ were ground and mixed in an
agate mortar. The degree of homogeneity was defined by the
duration of grinding and mixing: A : 5 hours and B : 10 minutes.
The powder mixture was pressed into a bar ($1 \times 4 \times 30mm^3$) at
about $6t/cm^2$, heated at 900°C for 20 hours and then furnace
cooled. Subsequent heat-treatment was made under high
presure oxygen (650°C, 40 atm. O_2).

Compositions of samples, preparation conditions and
results of X-ray analysis are listed in TABLE I.

All samples of $La_2Cu_{1+x}O_z$ were completely single-
phase, and had an orthorhombically distorted K_2NiF_4 type
structure. The effect of the degree of mixing was not
observed in the X-ray diffraction pattern. The chemical
analysis showed that the sample of La_2CuO_z with the degree

of homogeneity A, prepared by heating at 900°C in air had the composition of $2 : 1 : 4$ within the accuracy of 0.1% and was not oxygen-deficient.

RESULTS

The temperature dependence of the electrical resistivity of La-Cu-O compounds are shown in FIGs. 1 and 2. Stoichiometric and homogeneous La_2CuO_4 (#1) is semiconducting down to 4.2 K, as has been reported by many researchers.[3-5] A sharp drop in the electrical resistivity with the onset temperature of about 40 K is observed in La_2CuO_z with the degree of homogeneity B (#2). Two non-stoichiometric samples, $La_2Cu_{1.04}O_z$ and $La_2Cu_{0.96}O_z$ with the degree of homogeneity A (#3 and #4) are semiconducting, as shown in FIG. 1. Homogeneous non-stoichiometry does not bring about the superconducting transition.

After the heat treatment at 650°C under high pressure oxygen (40 atm. O_2), all samples studied (#5, #6, #7 and #8) show an anomaly at about 40 K in the resistivity vs. temperature curve (FIG. 2).

DISCUSSION

Though quantitative estimation has not been made, it is certain that a small fraction of the sample becomes superconducting below 40 K in inhomogeneous and in oxygen-treated La_2CuO_z.

We consider that the heat treatment under high pressure of oxygen introduces excess oxygen to the sample.

The calculation of the ion distribution in the K_2NiF_4 lattice of La-Cu-O compounds was made under following two assumptions.

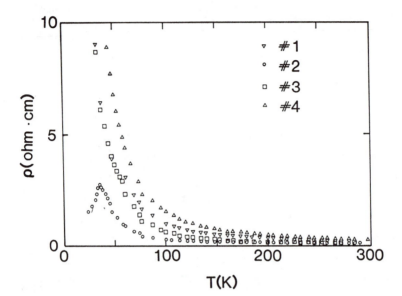

FIGURE 1.   Temperature dependence of the resistivity of
La-Cu-O compounds.

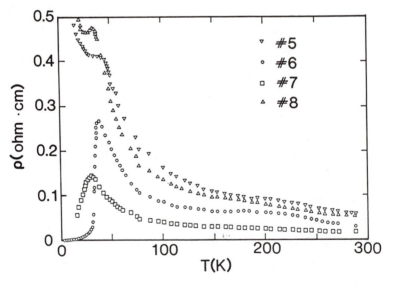

FIGURE 2.   Temperature dependence of the resistivity of
oxygen treated La-Cu-O compounds.

TABLE II.   Ion distributions on K and Ni sites in La-Cu-O compounds.

	La	Cu	O	K site	Ni site	F site
La_2CuO_4 (A)	La_2	Cu	O_4	La_2	Cu	O_4
O excess La_2CuO_z (A)	La_{2-2y}	Cu_{1-y}	O_4	La_{2-2y} Cu_{2y}	Cu_{1-3y} $\square_{3y}$	O_4
La excess region in La_2CuO_z (B)	La_2	Cu_{1-y}	O_4	La_2	Cu_{1-y} $\square_y$	O_4
Cu excess region in La_2CuO_z (B)	$La_{2-\frac{2}{3}y}$	$Cu_{1+\frac{2}{3}y}$	O_4	$La_{2-\frac{2}{3}y}$ $Cu_{\frac{2}{3}y}$	Cu	O_4

$\square$: vacancy

1) Ions do not occupy interstitial sites.

2) Vacancies exist only on the Ni sites.

The results are shown in TABLE II

It is noteworthy that the following two situations coexist in samples of both oxygen-excess and inhomogeneous La_2CuO_z.

1) Vacancies exist on the Ni sites.

2) A part of Cu ions occupies K sites together with La ions. These facts imply that the coexistence of these two situations has a certain relation to the occurrence of superconductivity in La_2CuO_z. It may be probable that superconductivity occurs in the disordered regions introduced by the inhomogeneity, where intergrowths of wrong layers such as $La_3Cu_2O_7$ or $La_4Cu_3O_{10}$ type exist.[6]

Although several reports[7-11] have been made on the occurrence of superconductivity in La_2CuO_z, the effects of the inhomogeneity, which is dependent on the degree of mixing and grain size of powders, were not checked in these studies.

Not only the introduction of excess oxygen by the heat treatment under high pressure of oxygen or plasma oxidation, but also the inhomogeneity must contribute to the occurrence of superconductivity in La_2CuO_z.

Careful examination of the condition for the occurrence of superconductivity in La_2CuO_4 is useful to clarify the mechanism of high-T_c superconductivity.

REFERENCES

1. K. Sekizawa, Y. Takano, H. Takigami, S. Tasaki and T. Inaba, Jpn. J. Appl. Phys. 26, L840 (1987).
2. Y. Takano, S. Tasaki, H. Takigami and K. Sekizawa, Proc. 18th Int. Conf. on Low Temp. Phys., Kyoto, 1987, Jpn. J. Appl. Phys. Suppl. 26-3, 1067 (1987).
3. R. J. Cava, R.B. van Dover, B. Batlogg and E.A. Rietman, Phys. Rev. Lett. 58, 408 (1987).
4. J.D. Jorgensen, K. Zhang and M.B. Brodsky, Phys. Rev. Lett. 58, 1028 (1987).
5. S. Uchida, S. Tanaka, Jpn. J. Appl. Phys. 26, L445 (1987).
6. A.H. Davis and R.J.D. Tilley, Nature (London) 326, 859 (1987).
7. J. Beille, R. Cabanel, C. Chaillout, B. Chevalier, G. Demazeau, F. Deslanded, J. Etourneau, P. Lajay, C. Michel, J. Provost, B. Raveau, A. Sulpice, J.L. Tholence and R. Tournier, C. R. Acad. Sci. Paris 304, 1097 (1987).
8. P.M. Grant, S.S.P. Parkin, V.Y. Lee, E.M. Engler, M.L. Ramirez, J.E. Vazquez, G. Lim, R.D. Jacowitz and R.L. Green, Phys. Rev. Lett. 58, 2482 (1987).
9. J.M. Tarascon, L.H. Green, B.G. Bagley, W.R. Mckinnon, P. Barboux and G.W. Hull. Novel Superconductivity, ed. S.A. Wolf et al. (Plenum Press, New York and London, 1987) p. 705.
10. S.A. Shaheen, N. Jisrawi, Y.H. Min, H. Zhen, L. Rabelsky, M. Croft, W.L. Mclean and S. Horn, ibid. p. 1083.
11. J.E. Schirber, J.F. Kwak, E.L. Venturini, B. Morosin, D.S. Ginley, W.S. Fu and R.J. Baughman, to be published.

MUON SPIN DEPOLARIZATION IN Gd- and EuBa$_2$Cu$_3$O$_x$

D. W. COOKE, R. L. HUTSON, R. S. KWOK, M. MAEZ, H.
REMPP, M. E. SCHILLACI, J. L. SMITH, AND J. O. WILLIS
Los Alamos National Laboratory, Los Alamos, NM 87545

R. L. LICHTI AND K. C. CHAN
Texas Tech University, Lubbock, TX 79409

C. BOEKEMA AND S. WEATHERSBY
San Jose State University, San Jose, CA 95192

J. OOSTENS
University of Cincinnati, Cincinnati, OH 45221

Abstract   Positive muon spin rotation (µSR) measurements
on Gd- and EuBa$_2$Cu$_3$O$_x$ (x ≈ 7) have been conducted in the
temperature interval 4 - 300K. For each sample, muons stop
both at grain boundaries and within the superconducting
grains. Measured   magnetic field penetration depths are
1550 and 1900Å for two specimens of GdBa$_2$Cu$_3$O$_x$, and
1350Å for EuBa$_2$Cu$_3$O$_x$.

Standard µSR techniques[1] were used to obtain positive muon
(µ$^+$) Gaussian depolarization rates Λ(T) for Eu- and
GdBa$_2$Cu$_3$O$_x$ in a 1 kOe transverse field. Also, zero-field data
were taken for the latter sample. Figures 1 and 2 show Λ(T)
and the corresponding muon precessional frequencies ν(T) for
a well-characterized polycrystalline sample of GdBa$_2$Cu$_3$O$_x$.
Two relaxation rates are observed for T<T$_c$, corresponding to
two distinct muon stopping regions. From the ν(T) data we
conclude that muons are thermalized in both normal and
superconducting volumes of the sample. Further, the measured
Fourier power and asymmetry data indicate that the
superconducting-to-normal volume ratio is 5.6. We suggest that

125

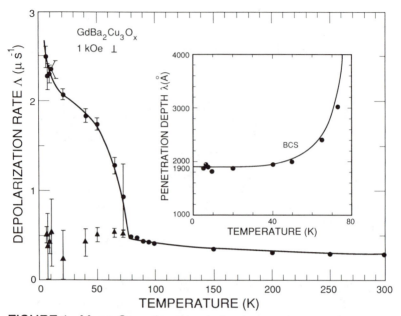

FIGURE 1. Muon Gaussian depolarization rates and penetration depth in GdBa$_3$Cu$_3$O$_x$. (▲ = normal, ● = supercond.)

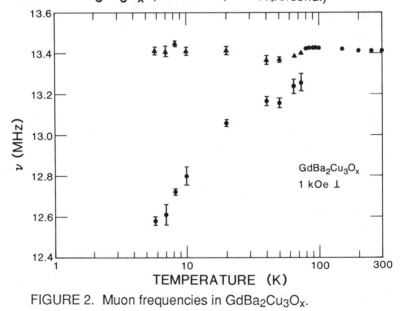

FIGURE 2. Muon frequencies in GdBa$_2$Cu$_3$O$_x$.

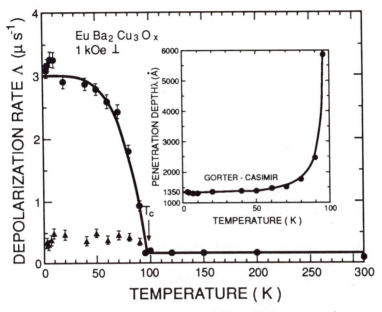

FIGURE 3. Muon Gaussian depolarization rates and penetration depth in $EuBa_2Cu_3O_x$.

85% of stopped muons reside within the superconducting grains and that 15% occupy grain boundaries. Similar results are obtained from polycrystalline $EuBa_2Cu_3O_x$ as shown in Figs. 3 and 4, and also from a second sample of $GdBa_2Cu_3O_x$ (not shown).

The magnetic field penetration depth $\lambda(T)$ is related to the field inhomogeneity, and thus $\Lambda(T)$, by the expression

$$< \mid \Delta H \mid^2> = (B\phi/4\pi\lambda^2)[1+(4\pi^2\lambda^2B)/\phi]^{-1} = 2\Lambda^2/\gamma_\mu^2 \quad (1)$$

where Λ is the μ^+ depolarization rate, γ_μ is the muon gyromagnetic ratio, λ is magnetic penetration depth, and ϕ is the flux quantum. By properly extrapolating the measured depolarization rates of Figs. 1 and 3 to zero Kelvin, and

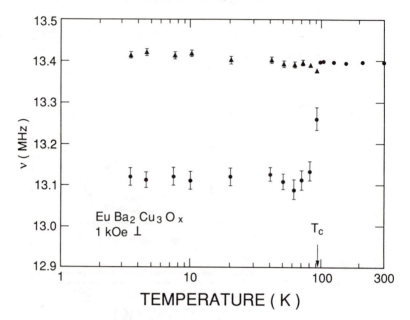

FIGURE 4. Muon frequencies in EuBa$_2$Cu$_3$O$_x$.

applying eq. (1), we find $\lambda(0)$ to be 1550 and 1900Å for two samples of GdBa$_2$Cu$_3$O$_x$, and 1350Å for EuBa$_2$Cu$_3$O$_x$. Moreover, we compute the temperature dependence of $\lambda(T)$ by assuming a phenomenological[2] expression $\lambda(T) = \lambda(0)[1 - (T/T_c)^4]^{-1/2}$ which we substitute into eq. (1), with the above measured values for $\lambda(0)$. These results are shown in the insets of Figs. 1 and 3. The solid lines represent the phenomenological expression calculated with the appropriate T_c's.

Zero-field μ^+ exponential relaxation rates and asymmetries for GdBa$_2$Cu$_3$O$_x$ are shown in Fig. 5. With decreasing temperature there is a drop in asymmetry, and a concomitant increase in Λ near T_c. Enhanced field inhomogeneity can explain the increase in Λ, but not the

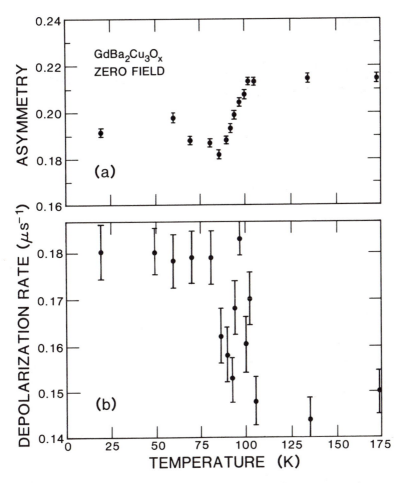

FIGURE 5. Muon exponential depolarization rates and asymmetries in GdBa$_2$Cu$_3$O$_x$ taken in zero field.

decrease in asymmetry; thus we offer no explanation for this puzzling result.

Work at Los Alamos is conducted under the aegis of the United States Department of Energy. Research support by Associated Western Universities, San Jose State Foundation, and the Robert A. Welch Foundation is acknowledged.

D. W. COOKE, ET AL.

REFERENCES

1. S. F. J. Cox, _J. Phys. C_, 20, 3187 (1987).
2. P. D. De Gennes, _Superconductivity of Metals and Alloys_ (Benjamin, New York, 1966), p. 26.

R. F. IMPENDANCE MEASUREMENTS ON YBa$_2$CU$_3$O$_7$

R. G. GOODRICH, C. Y. LEE, AND C. N. WATSON
Department of Physics and Astronomy
Louisiana State University
Baton Rouge, LA 70803-4001

Abstract R. F. impedance measurements were performed on YBa$_2$Cu$_3$O$_7$ and analyzed on the basis of
2-dimensional conduction.

INTRODUCTION

The electrical conductivity in the normal state (T > T$_c$) of
YBa$_2$Cu$_3$O$_7$ (YBC) is highly anisotropic. In YBC single
crystal studies Ong et al.[1] have shown that the conductivity
parallel to the planes just above T$_c$ is about 300 times the
conductivity perpendicular to the planes. Thus, it appears
that the superconductivity (SC) in these materials is mostly
confined to the Cu-O planes, and it is worthwhile to search
for special effects of a two-dimensional (2D) electron gas
that might be occurring.

The superconducting 2D electron gas is a
subject that has been investigated both theoretically and
experimentally for a number of years[2]. It exhibits a
different type of phase transition from zero resistance to a
resistive state than do three-dimensional superconductors,
due to bound fluxoid formation in two dimensions that does
not occur in three dimensions. In particular, there is a
transition from a system of bound fluxoid pairs that do not
cause coherent phase disruption of the SC electronic state

131

wavefunction to a system of free fluxoids that do disrupt the phase. This transition occurs below the mean field T_c for the material. The existence of this transition was first discussed by Kosterlitz and Thouless[3] and is known as the Kosterlitz-Thouless (KT) transition. The observed temperature dependence of the resistance below T_c in thin films of low-temperature superconductors agrees with most of the predictions of this theory.

Another observable effect of a 2D system of superconducting electrons is that they have a unique temperature dependence of fluctuations above T_c before they condense into the superconducting state. The theory of this effect was given by Aslamazov and Larkin (AL) and has been observed in thin film superconductors[4]. The KT transition and the AL fluctuations are related, in that they both depend on the same mean field transition temperature, T_o, with the KT theory applying at $T < T_o$ and AL theory for $T > T_o$.

EXPERIMENT

We have made 10 MHz complex impedance measurements on ceramic YBC samples having very sharp total transitions, $R_{90\%}$ to $R_{10\%} = 0.5$ K. The real part of the impedance at the transition is shown in Figure 1.

The central part of our observed transition is very sharp, but there is a rounding both on the high temperature side and on the low temperature side. One advantage of the technique we use to measure these transitions is that no electrical contacts are involved, because the measurements are made from the losses due to induced currents in the sample. The imaginary part of the impedance shows similar effects on both sides of the transition.

R. F. IMPEDANCE MEASUREMENTS ON $YBa_2Cu_3O_7$

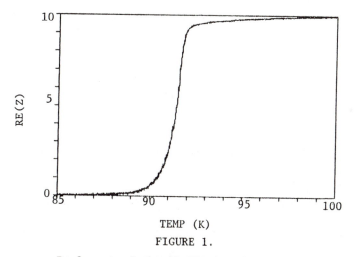

TEMP (K)

FIGURE 1.

Real part of the 10 MHz impedance (arbitrary units) of ceramic $YBa_2Cu_3O_7$ vs. temperature at 0.05 Gauss.

The measurements were taken in external magnetic fields from near zero (H_{ext} < 0.05 gauss) to 150 gauss. Measurements in fields as small as 1 gauss show a broadening of the low temperature tail on the transition. While there may be some broadening of the SC transition due to inhomogeneity in the ceramic samples, it is small because the main part of the transition is very sharp. We have made the assumption that the rounding on the high temperature side and the tail on the low temperature side are both due to 2D effects; we have analyzed our data with this assumption.

According to KT theory the resistance (Re(Z)) between the KT transition temperature, T_{KT}, and the mean field transition, T_o, should depend on temperature in the following way:

$$ln(R) \sim [(T_o - T)/(T - T_{KT})]^{1/2}.$$

133

R. G. GOODRICH, C. Y. LEE, AND C. N. WATSON

From AL theory, the high temperature side of the transition should obey:

$$1/R = 1/R_N + (1/R_o)*[T_o/(T - T_o)],$$

where R_N is the normal state resistance and R_o is a universal constant.

We have determined T_o by fitting the high temperature side of our data to AL theory. A least squares procedure for one, two, and three dimensional fits was done; the residual sum of squares for two dimensions was smaller by a factor of two than for one or three dimensions. Using this value of T_o a plot of $\ln(R)$ vs. $[(T_o-T)/((T-T_{KT})]^{1/2}$ for an applied field of 5 gauss is shown in Figure 2. As can be seen, the plot is linear, suggesting that 2D effects may be playing a role in this transition. From the data taken in other magnetic fields ranging from 0.05 to 200 gauss we find that the low-temperature side of the transition is greatly broadened, with an observable effect at applied fields > 1 gauss, and saturating near 150 gauss.

Each of these sets of data shows the same linear KT behavior with a different value of T_{KT}. The values of T_{KT} decrease smoothly with increased applied field and start to saturate above 20 gauss. In Figure 3. we plot the value

134

of the KT temperature obtained from these plots against applied magnetic field.

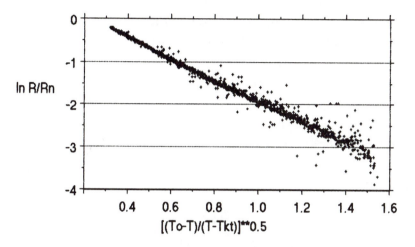

FIGURE 2.

Kosterlitz-Thouless plot for B = 5 Gauss and T_{KT} = 83.0 K.

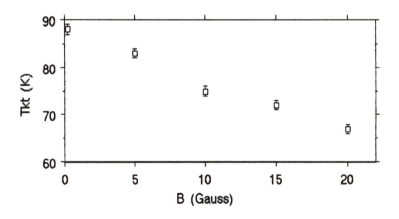

FIGURE 3.

Kosterlitz-Thouless temperature vs. applied magnetic field.

R. G. GOODRICH, C. Y. LEE, AND C. N. WATSON

CONCLUSIONS

The theory of rf interactions with a 2D gas is at best incomplete, and we know of no theory that completely describes the expected behavior in a magnetic field. We point out that there have been many reports of Josephson tunneling between grains in ceramic samples, and this could be the origin of the large field dependence. On the other hand, it is difficult to see how Josephson effects can account for the linear dependence of ln(R) on the KT expression for all applied fields.

REFERENCES

1.   N. P. Ong, S. J. Hagen, Z. Z. Wang, and T. W. Jing, Paper B2., this conference.
2.   P. Minnhagen, Rev. Mod. Phys.,59, 1001 (1987)
3.   J. M. Kosterlitz and D. J. Thouless, J. Phys. C 6, 1181 (1973)
4.   L. G. Aslamazov and A. I. Larkin, Fiz. Tverd. Tela. (Leningrad) 10, 1104 (1968) [Sov. Phys.- Solid State 10, 875 (1968)]

EPR OF HIGH-TEMPERATURE SUPERCONDUCTORS

AVETIK HARUTYUNYAN, LEONID GRIGORYAN, MIKAEL TER-MIKAELYAN
Institute for Physical Research, Armenian Academy of Sciences, Ashtarak, USSR 378410

Abstract A microwave absorption of the ceramic $REBa_2Cu_3O_{9-y}$ samples at low magnetic fields in a superconducting state is investigated. A resonant origin of the absorption is discussed.

INTRODUCTION

Studies of microwave absorption of the oxide superconductors by EPR spectroscopy revealed an extremely asymmetric signal near zero external magnetic field appearing just below T_c.[1-4] At present time there are a number of interpretations of the signal's origin (for a review see Ref 2).

EXPERIMENTAL

Ceramic samples of $REBa_2Cu_3O_{9-y}$ (RE=Y, Gd, Er or Ho) were prepared by solid-state reaction, and are characterized by resistive T_c=89-95 K, ΔT_c= 0.5-1.2 K.[2] The magnetic field derivative of the absorbed microwave power dP/dH vs H spectra were measured in RE-1306 or Bruker ESP-300 spectrometers (9.4 GHz). The Helmholtz coils were attached to the magnets of spectrometer to compensate for a residual field. The samples were carefully centered in a cavity. The points which didn't shift upon amplification of the signal were taken as zero values of dP/dH.

137

A.HARUTYUNYAN, L.GRIGORYAN, M.TER-MIKAELYAN

RESULTS AND DISCUSSION

Low-Field Signal

In **Fig.** 1 a typical absorption spectrum of zero field ($<$ 0.2 mT) cooled $REBa_2Cu_3O_{9-y}$ samples is shown at 77 K. The microwave power was $<0.10^{-6}$W, the 100 kHz modulation field amplitude <0.1 mT, and the external $H\perp H_{rf}$ ($H_{rf}$is the magnetic component of microwave field). It should be stressed that a change of polarity of the external H results in a mirror reflection of the signal with respect to the H axis (dotted line, Fig. 1). With an external $H/\!/ H_{rf}$ no signal was detected. This means that the sample selectively absorbs only one of the circular components of the linearly polarized microwave radiation. At very low available valu-

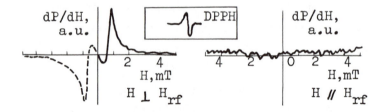

FIGURE 1. The spectra of $REBa_2Cu_3O_{9-y}$ and DPPH (inset) at 77K.

es of microwave power (ca. 10^{-6}W) an inhomogenuous saturation was observed, bringing about a deformation of the signal. We conclude that the signal is due to a resonant absorption of microwave power by the superconducting samples.

EPR OF HIGH TEMPERATURE SUPERCONDUCTORS

The shape of the signal should be Dysonian
when the diffusion time of the conduction electrons
is much shorter than the relaxation time. We
believe that the apparent low values of the resonant
field may be ascribed to a significant difference
between the externally applied H and the effective
field within the superconducting sample.   An
effective mass must be taken into account also.

It should be noted that at a fixed polarity of
the external field, the low-field extremum of the signal corresponds to a negative, and the high field one-to a positive value of dP/dH, while in
the case of common paramagnetic samples (e.g.DPPH)
this situation is reversed (Fig. 1). This may
indicate that the superconductor and DPPH
absorb   different circular components of the microwave field, i.e. the resonating magnetic moments in the superconducting sample and DPPH precess in
opposite directions. So it is natural to relate
this phenomenon to the strong diamagnetism of the
shielding supercurrents. However, the measurements
using   circularly polarized microwaves may be
needed to confirm this conclusion.

Specific Features of the Signal

We investigated factors which may strongly influence the signal (Fig. 2): magnetic history and
particle sizes of the samples; direction of field
sweep (i.e. sign of dH/dt, where t is time); amplitude of the modulation field. An effect of magnetic history is due to   flux trapping by the samples. Cooling of the sample below T_c, at a field

A.HARUTYUNYAN, L.GRIGORYAN, M.TER-MIKAELYAN

H_o influences the spectrum from $H_o=0.5-0.8$ mT up to $H_o=20$ mT. A comparison with the magnetization curve shows that 20 mT is the field where the magnetization of the sample reaches its maximum

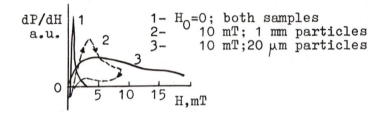

FIGURE 2.   Influence of different factors on the parameters of the signal.

value. An increase of H_o above 20 mT doesn't affect the signal any more. If a sample was cooled down to 77 K at fields not exceeding 2 mT, we managed to recover the signal characterizing the zero-field cooled sample by applying the compensating external magnetic field.

These, as well as some other phenomena related to microwave absorption of high-T_c ceramics at external magnetic field, are discussed in details in[2].

REFERENCES

1. A.R. Harutyunyan, L.S. Grigoryan, in Proc.Int. Conf. Low Temp. Phys., Budapest, 1987, p.31.
2. A.R. Harutyunyan, L.S.Grigoryan, A.V. Gevorkyan, A.S. Kuzanyan, Preprint of Institute for Physical Research   - 130, 1988.
3. K.W. Blazey, K.A. Müller, J.G. Bednorz, W. Berlinger, G. Amoretti, E. Bulaggiu, A. Vera, F. Matacotta., Phys. Rev B, 36, 7241 (1987).
4. V.I. Alexandrov, A.G. Badalyan, P.G. Baranov, V.S. Vikhnin et al. Pisma JETF, 47 (1988).

ANOMALOUS SUPERCONDUCTIVITY IN THE SYSTEM Y-Ba-Cu-Ag-O

E. GATTEF[a], E. VLAHOV[b], V. KOVACHEV[b], S. JAMBASOV[a], S. TINCHEV[c], AND M. TASLAKOV[c]

a) Higher Institute of Chemical Technology, Sofia-1156, Bulgaria.
b) Institute of Solid State Physics, Bulgarian Academy of Sciences, Sofia-1000, Bulgaria.
c) Institute of Physical Electronics, Bulgarian Academy of Sciences, Sofia-1000, Bulgaria.

Abstract A mixture of yttria, silver oxide, barium carbonate and copper oxides, fired in air at 960°C, provided two samples, each with two transitions in AC four-probe susceptibility measurements. The upper transitions are at T_c (midpoint) = 97.4 K (80% of sample) and 93.7 K (35%) respectively; the lower transitions occur around 83 K (20%) and 81 K (65%) respectively. X-ray diffraction shows the presence of two phases, D and S, with roughly consistent volume fractions.

INTRODUCTION

As is known from the temperature-oxygen concentration diagram of the YBa_2Cu_3-oxide,[1] it is difficult to synthesize single-phase ceramics in the Y-Ba-Cu oxide system.[2] If the composition is different from the Y:Ba:Cu = 1:2:3 ratio, several crystalline phases may appear.[3-7] For the multiple-phase materials an anomalous temperature variation of the resistivity is observed.[8-14] Anomalous superconductivity consists in the appearance of at least two superconducting transitions.[8] The interval of established onset temperatures of the second transition may be divided in two regions, namely, 80-90 K[8-10] and 60-65 K.[11-14]

The origin of the anomalously high superconductivity may be

141

attributed to a metal-semiconductor transition, or to the occurrence of a partial superconductivity in the ceramic specimens.[8] Somekh et al.[12] considered that the anomaly is due partly to the two-dimensionality of the structure, and Yamaya et al.[13] pointed out that the anomaly is probably due to a new phase, or to a mixing effect of the orthorhomic phase and a new phase. Nakazawa et al.[11] established that the second transition, around 60 K, indicates the existence of another orthorhomic phase. From the X-ray powder pattern they found the latter phase is appreciably different from the higher-T_c phase. However Goldfarb et al.[15] have reported a second superconducting transition in an X-ray single-phase specimen, and Narayan et al.[16] have found that the first transition, around 290 K, is due to a new phase, which epitaxially grows on the orthorhombic $T_c = 90$ K phase.

These findings suggest further investigations of the anomalous superconductivity. In this paper we report experimental results about the anomalous superconductivity in the Y-Ba-Cu-Ag oxide system.

SYNTHESIS AND AC SUSCEPTIBILITY MEASUREMENT

Two kinds of samples, D-23 and D-24, were prepared from a mixture of Y_2O_3, AgO, $BaCO_3$ and CuO, heated at 960°C in air, and cooled at different rates in an oxygen atmosphere. The electric resistivity (AC suceptibility) measurements were performed using a conventional four-probe AC technique.

The results of the electrical resistivity measurements for fresh samples are given in Table 1. Figure 1 presents the AC susceptibility-temperature dependence in a magnetic field of 0.38 Oe.

The main pecularities are the two transitions in the χ - T curves. This suggests the existence of two superconducting phases. The low-temperature transitions are very sharp (ΔT_c about 2-3 K). This confirms the existence of clearly separated phases. The ratio of the

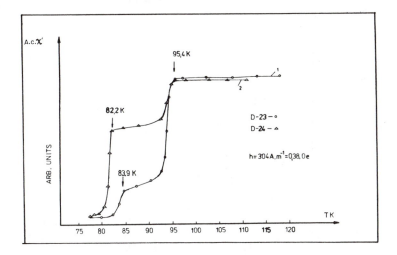

FIGURE 1. AC susceptibility versus temperature.

volume fractions of the high-temperature to the low-temperature phases may be estimated for D-23 as 80/20, and for D-24 as 35/65, provided that the low-temperature phase does not contribute to the susceptibility at high temperatures.

Table 1. Critical Temperatures for superconductivity in Y-Ba-Cu-O doped with AgO.

Sample	T_c(onset) /K	T_c(mid-point) /K	T_c(R=0) /K	ΔT_c (90%-10%)/K
D-23	107	97.4	95.8	0.9
D-24	102	93.7	92.8	1.4

143

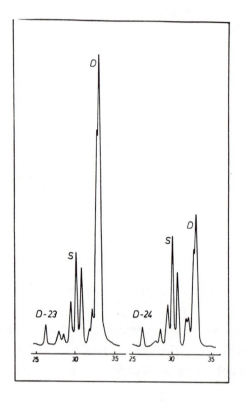

FIGURE 2. X-ray patterns in 2Θ scale; CuKα radiation.

X-RAY DIFFRACTION

The X-ray diffraction patterns (Fig. 2) show that the samples contain two crystalline phases. These patterns are similar to those presented in Ref. (3). The first phase, named the D-phase, corresponds to the compound Y_2BaCuO_5. The quantitative X-ray powder diffraction analyses gives for the volume fraction ratio of D-phase to S-phase, 76/24 for D-23, and 54/46 for D-24. The latter results correlate well with the results obtained from the AC susceptibility measurements.

ANOMALOUS T_c IN THE Y-Ba-Cu-Ag-O SYSTEM

CONCLUSION

The results of the present study confirm that in the multiphase oxide materials one can obtain anomalously high temperature superconductivity. When the volume concentration of S-phase is increased, the onset temperature of the second transition is decreased.

ACKNOWLEDGEMENT

This investigation has been completed with the financial support of the Committee of Science under contract No. 537.

REFERENCES

1. J. Hauck, K. Buckman, F. Zucht, Z. Phys. B., Condensed Matter, 67 (1987) 299.
2. K. Takagi, K. Miyauchi, Y. Ito, T. Aida, H. Hasegawa, U. Kawabe, Jpn. J. Appl. Phys., 26 (1987) L699.
3. T. Hioki, A. Oota, M. Ohkubo, T. Noritake, J. Kawamoto, O. Kamigato, Jpn. J. Appl. Phys., 26 (1987) L873.
4. Y. Nakazawa, M. Ishikawa, T. Takabatake, K. Koda, K. Terakura, Jpn. J. Appl. Phys., 26 (1987) L796.
5. C. Rao, P. Cauguly, A. Raychaudhury, R. Ram, K. Sreedar, Nature, 326 (1987) 30 April.
6. D. Hinks, L. Soderholm, D. Capone, J. Jorgensen, I. Schuller, C. Segre, K. Zhang, J. Grace, Appl. Phys. Lett., 50 (1987) 1688.
7. Y. Oda, T. Kohara, I. Nakada, H. Fujita, T. Kaneko, H. Toyoda, E. Sakagami, K. Asayama, Jpn. J. Appl. Phys., 26 (1987) L807.
8. Y. Takagi, R. Liang, Y. Inagyma, T. Nakamura, Jpn. J. Appl. Phys., 26 (1987) L1266.
10. Y. Nakazawa, M. Ishikawa, T. Takabatake, H. Takeya, T. Shibuya, K. Terakura, F. Takei, Jpn. J. Appl. Phys., 26 (1987) L682.
11. R. Somekh, M. Blamire, Z. Barber, K. Butter, J. James, G. Morris, E. Tomlinson, A. Schwarzenberger, W. Stobs, J. Evets, Nature, 326 (1987) 857.

EXPERIMENTAL

The compound $YBa_2Cu_3O_7$ was prepared by the method des-cribed earlier[1]. Bulk thermal expansion measurements were carried out using a Model LKB 3185 fused quartz pushrod-type dilatometer. The sample, in the form of a pressed, sintered, cylindrical pellet of stoichiometry $YBa_2Cu_3O_7$ and bulk den-sity of $\sim$65% theoretical density, was used. The measurements were made on two different pellets of same stoichiometry and bulk density with a heating rate of 5°/min in the range 298-1173 K in air. The oxygen content of the sample was deter-mined at different temperatures from the micro-TG data ob-tained under identical conditions. A Shimadzu TG instrument, mo-del TGC-31 was used for this purpose.

RESULTS AND DISCUSSION

The dilatometric data are shown in Fig.1. The variation of per-cent linear expansion ($\Delta L/L \times 100$) with temperature in the range 298-1173 K in air is expressed by the following least-squares fitted equation.

$$\Delta L/L \times 100 = 4.402 \times 10^{-4}(T-298) + 4.029 \times 10^{-6}(T-298)^2$$
$$-2.908 \times 10^{-9}(T-298)^3 \qquad (1)$$

The temperature dependence of coefficient of average linear expansion ($\alpha_1 = \Delta L/L\Delta T$) derived from Eq. (1) above is represented by Eq.(2) below

$$\alpha_1 = 4.402 \times 10^{-6} + 4.029 \times 10^{-8}(T-298) - 2.908 \times 10^{-11}(T-298)^2$$
$$(2)$$

The values of α_1 calculated at different temperatures using Eq. (2) are given in Table I along with the stoichiometry of the sample determined on the basis of oxygen loss observed on micro-TG during heating.

BULK THERMAL EXPANSION OF $YBa_2Cu_3O_7$

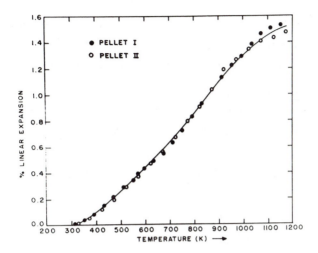

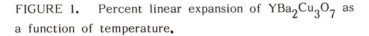

FIGURE 1. Percent linear expansion of $YBa_2Cu_3O_7$ as a function of temperature.

As can be seen from Table I, the oxygen content of the sample decreases with temperature above 600 K, while the coefficient of bulk thermal expansion (α_1) increases up to 1000 K, and thereafter it shows a gradual decrease. It is reported in the literature[2] that as the oxygen content decreases, the orthorhombic phase of $YBa_2Cu_3O_7$ changes to a tetragonal phase. This transition is known to occur around 900 K in air and is accompanied with a slight decrease in the unit cell volume followed by an increase. The continuous decrease observed in the bulk thermal expansion coefficient above 1000 K is attributed to the porosity changes. The evolution of oxygen on heating the sample in air increases the void space along the grain boundaries, which causes the bulk porosity to increase. But at sufficiently high temperatures (>1000 K), sintering phenomenon predominates which decreases the bulk porosity. The decrease in bulk porosity results in the decrease of bulk volume, and hence the coefficient

149

of bulk thermal expansion decreases. However, the average value of the bulk thermal expansion coefficient in the entire range of investigation is quite high $(\alpha_1 = 17.4 \times 10^{-6} K^{-1})$, which suggests that the oxygen ions in $YBa_2Cu_3O_7$ are loosely bound.

TABLE I   Dilatometric data for $YBa_2Cu_3O_7$

Temperature (K)	Stoichiometry	Coeff. of av. linear exp. $\alpha_1 \times 10^6\ K^{-1}$
400	$YBa_2Cu_3O_7$	8.21
500	$YBa_2Cu_3O_7$	11.36
600	$YBa_2Cu_3O_7$	13,92
700	$YBa_2Cu_3O_{6.97}$	15.91
800	$YBa_2Cu_3O_{6.90}$	17.31
900	$YBa_2Cu_3O_{6.76}$	18.13
1000	$YBa_2Cu_3O_{6.65}$	18.38
1100	$YBa_2Cu_3O_{6.55}$	18.04
1173	$YBa_2Cu_3O_{6.49}$	17.40

REFERENCES

1. A.C. Momin, M.D. Mathews, V.S. Jakkal, I.K. Gopalakrishnan and R.M. Iyer, Solid State Commun., 64, 329 (1987)

2. P.K. Gallagher, H.M. O'Bryan, S.A. Sunshine and D.W. Murphy, Mater. Res. Bull., 22, 995 (1987)

EXPERIMENTS ON THE GROWTH OF Y-Ba-Cu OXIDE SINGLE CRYSTALS

IN-GANN CHEN and D.M. STEFANESCU
Department of Metallurgical and Materials Engineering,
The University of Alabama, P.O. Box G, Tuscaloosa, Alabama 35487

Abstract   A Bridgman-type directional solidification furnace was used to grow $YBa_2Cu_3O_x$ single crystals. Relatively large (maximum size about 2 mm x 2 mm x 0.01 mm) single crystals have been grown by this method. The chemical composition and crystallographic structures of these crystals were examined by SEM/EDS, and X-Ray diffractometry techniques. A possible growth mechanism of these crystals is discussed, based on experimental observations. It is proposed that a liquid-solid-gas reaction is involved in the growth of $YBa_2Cu_3O_x$ single crystals from off-stoichiometric 123 compositions in the temperature range of about 950 to 1000 °C.

INTRODUCTION

Although polycrystalline superconductive oxides can be prepared by relatively simple processes, such as powder-calcination or co-precipitation methods, the growth of large $YBa_2Cu_3O_x$ single crystals is still not well developed due to practical difficulties. It has been shown that $YBa_2Cu_3O_x$ decomposes through a peritectic transformation near 1020 °C in air[1], which prevents congruent growth of the $YBa_2Cu_3O_x$ single crystal directly from the 123 stoichiometric melt. Takekawa and Iyi[2] and Roth et al.[3] have suggested a preliminary ternary Y_2O_3-BaO-CuO phase diagram. A partially melted region is observed in the CuO-rich corner in the temperature range of about 950 to 1000 °C. These results indicate that an off-stoichiometry quaternary eutectic zone may exist at about 950 °C. This melt zone will allow the growth of large 123 single crystals directly from the melt. A number of reports have shown that relatively large

151

$YBa_2Cu_3O_x$ single crystals can be grown from these off-stoichiometric (Cu and/or Ba rich) melts[2,4-6]. This report describes experimental studies in a Bridgman-type directional solidification furnace, in an attempt to grow large-size $YBa_2Cu_3O_x$ single crystals from the off-123 partial melt. Relatively large (maximum size about 2 mm x 2 mm x 0.01 mm) single crystals have been grown by this method.

EXPERIMENTAL PROCEDURE

A Bridgman-type directional solidification furnace, shown schematically in Fig. 1, was used to grow $YBa_2Cu_3O_x$ single crystals directly from the oxide melt. This setup allows the control of the oxygen pressure in the crucible during the crystal growth process. A powder mixture of 123 + BaO + CuO with the chemical stoichiometry ratio of 1:3:9

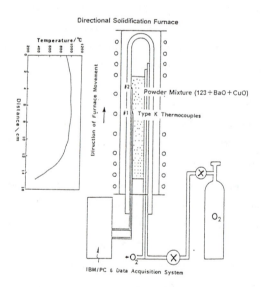

Figure 1. Tube crucible inside the Bridgman-type directional solidification furnace, and the temperature profile of this furnace.

(Y:Ba:Cu) was loaded into the Al_2O_3 tube crucible. Two type-K thermo-couples and two small ceramic tubes were also installed inside the Al_2O_3 crucible, as shown in Fig. 1. A typical temperature profile along the longitudinal direction is shown in the insert of Fig. 1. The crucible was then installed with the closed part upwards, to allow oxygen flow through the inner ceramic tubes. Pure oxygen flow was maintained during the course of the experiment. The bottom area of the powder mixture was kept below 700 ^{o}C, to prevent flow of the molten oxide out of the crucible. The temperature was monitored through a data acquisition system with an IBM/PC computer. The temperature was increased to about 1010 ^{o}C, and held for about 20 min., before translating the furnace in the upward direction at a rate of 0.1 mm/min. Fig. 2 shows the temperature profile for two thermocouples positioned 3.1 cm apart. The cooling rates at the locations of the two thermocouples were calculated to be about 10 and 60 ^{o}C/hr respectively.

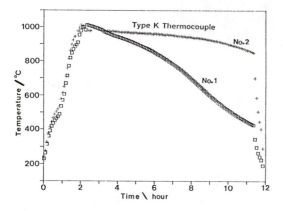

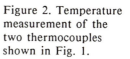

Figure 2. Temperature measurement of the two thermocouples shown in Fig. 1.

RESULTS AND DISCUSSION

The chemical composition and crystallographic structure of the sample after breaking the crucible were examined by SEM/EDS (X-Ray Energy Dispersive Spectrometry), and X-Ray diffractometry techniques. Fig. 3

the crystal growth mechanism. The fast growth rate in the a and b directions, to form plate-type crystals, is believed to be due to the strongly anisotropic crystal structure of the $YBa_2Cu_3O_x$ compound.

In the area near the thermocouple No. 2, where a higher temperature existed than for No. 1 thermocouple, needle-type crystals were observed, as shown in Fig. 6. Based on SEM/EDS and X-Ray diffractometry analyses as described above, the needle-type crystal is believed to be the 211 phase, which is consistent with the phase diagram, which shows that only the 211 phase is stable at high temperature.

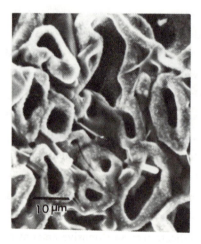 

Figure 5. SEM micrograph of the porous subsurface beneath the plate-type crystals.

Figure 6. SEM micrograph of needle-type single crystals.

GROWTH OF Y-Ba-Cu OXIDE SINGLE CRYSTALS

CONCLUSIONS

Relatively large, plate-type $YBa_2Cu_3O_x$ single crystals can be grown in a directional solidification furnace from an off-stoichiometry powder mixture in the temperature range of about 950 to 1000 ^{O}C. Needle-type 211 single crystals can be grown by the same method at higher temperatures. It is believed that larger crystals may be grown by further modification of the chemical composition, oxygen partial pressure, and operation temperature. A possible growth mechanism with more complicated liquid-solid-gas reaction is proposed to account for the crystal growth.

ACKNOWLEDGEMENTS

This work has been supported by the School of Mines and Energy Development, by the College of Engineering at the University of Alabama, and by NASA Grant No. NCC8-3.

REFERENCES

1. J. Hauck, K. Bickmann, and F. Zucht, J. Mater. Res. 2(6), 762(1987).

2. S. Takekawa and N. Iyi, Jpn. J. Appl. Phys., 26 L851(1987).

3. R.S. Roth, K.L. Davis, and J.R. Dennis, Adv. Ceram. Mater. 2, 303(1987).

4. D.L. Kaiser, F. Holtzberg, B.A. Scott, and T.R. McGuire, Appl. Phys. Lett. 51, 1040(1987).

5. M.A. Damento, K.A. Gschneidner, Jr., and R.W. McCallum, Appl. Phys. Lett., 51, 690(1987).

6. H. Katayama-Yoshida, Y. Okabe, T. Takahashi, T. Sasaki, T. Hirooka, T. Suzuki, T. Ciszek, and S.K. Deb, Jpn. J. Appl. Phys., 26 L2007(1987).

$0.0 \leq x \leq 3.0$) were prepared by conventional ceramic powder methods from stoichiometric mixtures of fine powders of their corresponding oxides. The resulting mixtures were pre-fired at 925°C in air for 8 hrs. prior to the sinter/anneal process. Cylindrical pellets of each of these materials were pressed, sintered at 925°C and annealed at 500°C, while keeping the entire firing schedule under controlled oxygen or ozone atmospheres.[4,5] Details of the processing procedure and the experimental set-up can be found elsewhere.[2-4]

Four-probe ac electrical resistance versus temperature measurements and x-ray diffraction studies were performed on pellets of each sample. Thermogravimetric analysis (TGA) was performed to determine the oxygen content[6] of each sample in flowing argon/hydrogen under identical conditions, while other samples were heated in argon only.

RESULTS AND DISCUSSION

Comparisons of the effect of processing under oxygen and ozone atmospheres, for typical normalized electrical resistance versus temperature data, are shown in Figures 1-3. Some of the characteristics noted previously,[4] such as a narrowing of the widths of the transitions with ozone processing, are apparent. (ΔT_c is defined as [10% - 90%] of the transition). The metal-substituted materials generally exhibit a decrease in T_c (zero-resistance) with increased substitution, yet most samples also exhibit similar trends with changes in atmospheric processing. TGA results for Y-Ba-Cu-O material heated in argon and analyzed under flowing oxygen, ozone and ambient air, shown in Figure 4, indicate that oxygen content is increased by ozone-treatment. X-ray diffraction studies revealed the orthorhombic $YBa_2Cu_3O_z$ phase to be the primary constitutent in all samples, although minor amounts of CuO and $BaCuO_2$ impurities were also present.

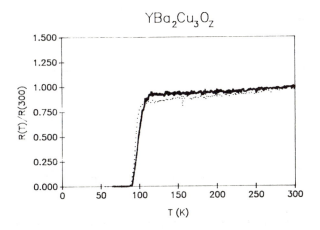

FIGURE 1. Normalized ac resistance versus temperature for $YBa_2Cu_3O_z$, where the solid line corresponds to the oxygen-processed sample, and the dotted line corresponds to the ozone-processed sample.

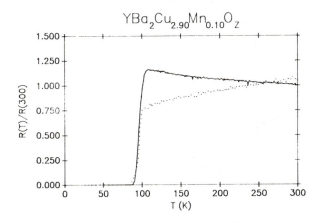

FIGURE 2. Normalized ac resistance versus temperature for $YBa_2Cu_{2.90}Mn_{0.10}O_z$, where the solid line corresponds to the oxygen-processed sample and the dotted line corresponds to the ozone-processed sample.

161

5. M. A. Maginnis, D. A. Stanley, D. A. Robinson, U. S. Patent Pending, 1987.
6. P. K. Gallagher, H. M. O'Bryan, S. A. Sunshine, D. W. Murphy, <u>Mat. Res. Bull.</u>, <u>22</u>, 995 (1987).

AN APPARATUS FOR IN-SITU MAGNETIC SUSCEPTIBILITY, THERMO GRAVIMETRIC ANALYSIS, GAS ABSORPTION, ETC. MEASUREMENTS

L.N. MULAY AND MARK KLEMKOWSKY
Materials Science and Engineering Dept., 136 MRL
Building, Pennsylvania State University, University Park,
PA 16802 U.S.A.

Abstract. A sensitive micro Faraday balance for magnetic measurements on superconductors is described. This apparatus can be used also for TGA, gas absorption/ adsorption and high pressure measurements on various materials. Typical applications are described.

INTRODUCTION

The prolific literature on the high T_c-oxide materials, including the compositions containing yttrium, bismuth, thallium, etc. shows that magnetic measurements were carried out mostly with the SQUID technique. The intrinsic cost of the sophisticated SQUID magnetometer as well as its operational cost, based on the use of liquid helium, is indeed very high. In addition it has limitations with regard to cycling of the steady magnetic field (H) and measurement temperature. In contrast, the well-known Faraday balance is less expensive with regard to its assembly and operation. In this paper we describe the construction of a sensitive Faraday microbalance (that is a force magnetometer) for measurement of the very weak diamagnetic susceptibility (χ), which is about $- 1 \times 10^{-7}$ cm^3g^{-1}. This is often referred to as the cgs unit or simply emu. This χ value multiplied by 12.56×10^{-3} gives the S.I.value in m^3kg^{-1}. The applied field (H in Oersteds) times 79.6 gives the S.I. unit, namely Am^{-1}. In addition we present general applications of the Faraday balance to in-situ studies of absorption/adsorption of gases, TGA and high-pressure measurements.

L. N. MULAY AND M. KLEMKOWSKY

PRINCIPLE OF THE FARADAY BALANCE

Jeans[1] gives a detailed mathematical theory for the force experienced by a body in a field (H) with a gradient dH/dx. From this theory, the following simple equation can be derived for the force measured in terms of the change in weight (ΔW_s), which a sample (s) with mass (M_s) experiences when the field is turned on.

$$\Delta W_s = m_s\, \chi_s\, H \cdot dH/dx$$

For quick susceptibility measurements a reference material of known susceptibility is employed. Denoting the corresponding reference parameters by the subscript r and eliminating $H \cdot dH/dx$ from the force equations for the sample and reference, the susceptibility for the sample (χ_s) is simply derived from

$$\chi_s = (\Delta W_s\ m_r)\chi_r/\Delta W_r m_s$$

The above relative method is described in detail by Mulay[2].

INSTRUMENTATION OF THE FARADAY MICROBALANCE

A sketch of the entire system is shown in Figure 1. The electro-microbalance with a load capacity of 100g and a sensitivity of 1 microgram was supplied by the Cahn Instrument Co., Cerritos, CA (Model R.G. 2,000). The sample holder was suspended from the weighing side of the electrobalance with a straight tungsten wire, at the end of which an ultrapure gold chain was attached. An electromagnet with 12" poles and a D.C. power supply (from Varian Associates, Palo Alto, CA) was used to generate steady fields (H) up to 18 kOe in a pole gap of 1.5". The electro magnet was placed on tracks so that it could be moved out of the way for changing the sample in a quartz bucket (0.25" ID and 0.5" tall). The entire Heli-Tran (Refrigerator) Cryostat (8 - 300 K, made by Air Products Co., Allentown, PA) could be removed when necessary so that a furnace (going up to 700°C) could be slipped around the quartz sample chamber. This feature allowed the use of the microbalance system for TGA studies.

The gradient field (dH/dx) was produced by a pair of "Lewis coils", connected to a separate bipolar D.C. power supply.   Thus the steady field (H) and the gradient (dH/dx) could be varied independently.   This advantage does not exist with slanted pole-tips of special contour, because, when H is varied dH/dx also changes[2].   Other advantages are described by Lewis (see ref. 138 given under ref. 2).

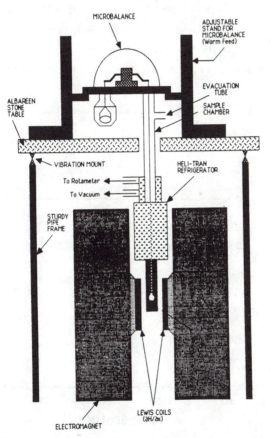

FIGURE 1.   Sketch of the Faraday Microbalance, showing major components.

L.N. MULAY AND M. KLEMKOWSKY

MAGNETIC SUSCEPTIBILITY MEASUREMENTS

The sample chamber is first evacuated up to 10^{-6} Torr using a Turbo Vacuum pump. A small amount of helium gas is then admitted so as to establish a thermal contact between the sample and the cryostat. Measurement of magnetic susceptibility of superconductors by the Faraday technique requires a number of precautions; for instance, the steady field (H) and the gradient (dH/dx) must be kept well below the critical field (H_{c1}). Since H_{c1} of about 300 Oe was reported for the $YBa_2Cu_3O_{7-\delta}$ compound the measurements of the susceptibility (χ in emu/g) were measured with H < 100 Oe and a small gradient. Typical results for the above compound (with $\delta = 0.05$) as a function of temperature are shown in Figure 2.

The observed $T_c \sim 95$ K was confirmed by the relative A.C. susceptibility measurements using an apparatus described by Cao and Mulay[3].

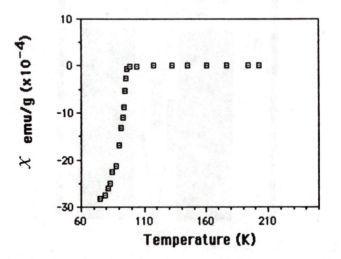

FIGURE 2. Magnetic susceptibility of $YBa_2Cu_3O_{7-\delta}$.

MAGNETIC SUSCEPTIBILITY--TGA APPARATUS

IN-SITU STUDIES ON GAS ABSORPTION

As a prelude to quantitative measurement on the uptake (absorption) of oxygen gas by the yttrium compound, air was admitted to the sample chamber and the system was allowed to reach equilibrium. The susceptibility measurements were then carried out in a routine way[2]. We were able to qualitatively verify that as the total oxygen content (7-δ) decreased from about 6.95 to 6.5, the T_c also seemed to lower from ~ 95 K to ~ 40 K. Our exploratory studies on the in-situ absorption studies carried out with the microbalance combined with the magnetic measurements show the potential of the system for such "in-situ" studies on oxygen deficient materials.

TGA MEASUREMENTS

The well known thermal decomposition shown below
$$CuSO_4:5H_2O \rightarrow CuSO_4:1H_2O + 4H_2O$$
has been verified with the Faraday balance. In this case, the cryostat is replaced by a tube-furnace wound noninductively with platinum wire. A temperature of 400°C was maintained during decomposition. The susceptibility of 1130×10^{-6} emu/mole was found for $CuSO_4:5H_2O$. The decomposition product $CuSO_4:1H_2O$ showed a susceptibility of 1178×10^{-6} emu/mole. The difference ($\Delta\chi = -48 \times 10^{-6}$ emu/mole) corresponds well with the loss of $4H_2O$ $[(4)(-12.9 \times 10^{-6})$emu/mole]. The agreement is even better when a correction for the loss in the paramagnetism of the Cu^{2+} ion at a temperature of 44°C is taken into account. The loss of $4H_2O$ was also verified directly with the microbalance. The above studies are encouraging indeed for combined in-situ TGA on various superconducting materials.

169

L. N. MULAY AND M. KLEMKOWSKY

PHYSISORPTION OF O_2 SHOWING EVIDENCE FOR O_4 TYPE SPECIES

Oxygen (O_2) which is paramagnetic with two unpaired spins $(^3\Sigma$ state) was shown to dimerize to a "diamagnetic" "O_4" type species, as predicted by Professor Pauling[4]. This was done by studying the physisorption of O_2 on high surface area γ - Al_2O_3 and concurrently measuring its magnetic susceptibility. The susceptibility 400 x 10^{-6} emu/g decreased to about 275 x 10^{-6} emu/g at 77 K when O_2 was physisorbed on the substrate. Mulay and Boudreaux[5] have outlined this work briefly; reference should be made to publications by Mulay et al cited therein for further details.

HIGH-PRESSURE STUDIES ON SmS

Mulay and coworkers[6] have successfully used the Faraday balance to elucidate the insulator to metal transition in SmS and doped SmS. A Cu-Be high pressure cell weighing about 60 g was used to measure the changes in susceptibility of SmS up to 6 K bars. Sm^{2+} was shown to change over almost to the Sm^{3+} state.

References

1. J.H. Jeans, "The Mathematical Theory of Electricity and Magnetism", Cambridge University Press, London, 1927.
2. L.N. Mulay, Chapter VII in Physical Methods of Chemistry, Vol I Part IV, A. Weissberger and B.W. Rossiter, Eds. John Wiley and Sons, Inc., New York, NY (1972).
3. Wenwu Cao, L.N. Mulay, et al. Mat. Sci. and Engr. 100, L11 (1988).
4. L. Pauling, "Nature of the Chemical Bond", Cornell University Press, Ithaca, NY (1960).
5. L.N. Mulay and E.A. Boudreaux, "Theory and Applications of Molecular Paramagnetism", and "Theory and Applications of Molecular Diamagnetism", John Wiley and Sons, Inc. New York (1976).
6. C. Cordero-Montalvo, K. Vedam and L.N. Mulay in "High Pressure Low Temperature Physics", C.W. Chu and J.A. Woolam, Eds., Plenum Press, New York (1978).

DOPING DIRECTED AT THE OXYGEN SITES IN $Y_1Ba_2Cu_3O_{7-\delta}$: THE EFFECT OF SULPHUR, FLUORINE, AND CHLORINE

D. E. FARRELL and D. BOYNE
Case Western Reserve University, Cleveland, OH 44106

N. P. BANSI
NASA-Lewis Research Center, Cleveland, OH 44135

Abstract. We have investigated the effect of three dopants directed at the oxygen sites in $Y_1Ba_2Cu_3O_{7-\delta}$: sulphur, fluorine and chlorine. Single-phase material has been obtained up to a (nominal) replacement of the order of 1% of the oxygen. Although the lattice parameters are essentially unchanged, all dopants raise T_c (very slightly), sharpen the resistive transition, reduce the normal state resistivity, and very substantially increase the (magnetically determined) fraction of the material that is superconducting. All these results differ qualitatively from those obtained with dopants directed at other locations in the 123 structure, and it is suggested that small additions of sulphur, fluorine or chlorine may help to stabilize the ideal 123 stoichiometry.

MÖSSBAUER DEBYE-WALLER FACTORS IN HIGH-Tc SUPERCONDUCTORS[1]

I. ZITKOWSKY, R. N. ENZWEILER, P. BOOLCHAND, and W. HUFF
University of Cincinnati, Cincinnati, OH 45221, USA

R. L. MENG, P. H. HOR and C. W. CHU
University of Houston, Houston, TX 77004, USA

C. Y. HUANG
Lockheed Palo Alto Research Center Laboratory, Palo Alto, CA, USA.

Abstract. T-dependent ^{151}Eu and ^{119}Sn Mössbauer Debye-Waller factors f(T) in $EuBa_2Cu_{2.98}Sn_{0.02}O_{7-\delta}$ samples have been measured in the range 10K < T < 300K. Two qualitatively different f(T) variations are observed. ^{151}Eu f(T)'s display a normal behavior characteristic of a θ_D = 280(5) K over the T-range examined.[2] On the other hand, ^{119}Sn f(T)'s display evidence of a normal behavior in the range 120 K < T < 300 K with substantial softening[2] taking place at T < 120 K. The Sn dopant, we suggest, replaces largely Cu(1) chain sites, and the f(T) anomaly is the signature of a softening of the chain vibrational modes as a precursor to the onset of superconductivity at T_c ~80 K. These results suggest that the phonons of the chains probably play an important role in electron pairing in these novel materials.

1) Supported by NSF Grant DMR-85-21005.
2) P. Boolchand et al., Solid State Commun. **63**, 521 (1987) and unpublished.

samples which we use is the following. Appropriate amounts
of Tl_2O_3, CaO and Ba-Cu oxide (depending on the desired
stoichiometry) were completely mixed and ground, and pressed
into a pellet with a diameter of 7 mm and a thickness of 1-2
mm. The pellet was then put into a tube furnace which had
been heated to 880-910 °C, and was heated for 2-5 minutes in
flowing oxygen, followed by furnace cooling to below 200 °C.
Quenching in air from 900 °C to room temperature depresses
T_c only slightly. A number of samples with different
stoichiometry, including a series of samples with nominal
compositions of $Tl_2Ca_yBaCu_3O_{7+y+x}$ with y = 1, 1.5, 2, 3 and
4, were prepared. Samples with nominal compositions of
$Tl_{1.86}CaBaCu_3O_{7.3+x}$, $Tl_2CaBa_2Cu_2O_{8+x}$ and $Tl_2Ca_2Ba_2Cu_3O_{10+x}$
were also synthesized. All samples readily levitate over a
magnetic field. The strong Meissner effect observed is due
to both a large volume fraction of superconductive phase and
a higher transition temperature. Furnace-cooling improves
slightly the superconducting behavior of the samples. This
indicates that the Tl-Ca-Ba-Cu-O superconducting compound is
stable over a relatively large range of temperatures, and
thus the Tl-Ca-Ba-Cu-O superconductor should be of
importance in many applications.

RESISTANCE

Fig. 1 shows the resistance-temperature variation for a nominal $Tl_{2.2}Ca_2Ba_2Cu_3O_{10.3+x}$ sample. This sample has an onset

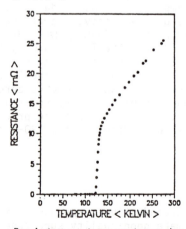

Figure 1   Resistance-temperature dependence of a
nominal $Tl_{2.2}Ca_2Ba_2Cu_3O_{10.3+x}$ sample.

temperature near 140 K, midpoint of 127 K, and zero
resistance temperature at 122 K. The highest zero
resistance temperature reported so far for the Tl-Ca-Ba-Cu-O
system is 125 K [10].

MAGNETIZATION

Magnetization measurements on the Tl-Ca-Ba-Cu-O samples were
performed utilizing a SQUID magnetometer manufactured by BTI
Corp., San Diego, CA [9]. Fig. 2 shows DC magnetization
(field cooled and zero field cooled) as a function of
temperature for an applied field of 1 mT for a nominal
$Tl_2Ca_4BaCu_3O_{11+X}$ sample. The insert of Fig. 2 shows data
traces for the same sample, and also for a well-prepared
$EuBa_2Cu_3O_{7-X}$ sample, where the vertical axis represents the
dX"/dH signal of an EPR spectrometer. As is seen from the

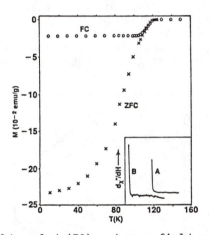

Fig. 2 Field cooled (FC) and zero field cooled (ZFC)
magnetization as a function of temperature for a
DC field of 1 mT for a nominal $Tl_2Ca_4BaCu_3O_{11+X}$
sample. The two data traces in the insert
illustrate the sharp onset of superconductivity
observed by the microwave technique as described
in the text. Sample A is $Tl_2Ca_4BaCu_3O_{11+X}$ with
onset temperature of 118.3 K, and for comparison,
data for a sample of $EuBa_2Cu_3O_{7-X}$ (sample B)
with onset temperature of 94.4 K is plotted.
Note that the difference in onset temperatures is
23.9 K.

insert, the onset temperature for the sample A
($Tl_2Ca_4BaCu_3O_{11+X}$) is 118.3 K, 23.9 K higher than that of
the sample B ($EuBa_2Cu_3O_{7-X}$, whose onset temperature is 94.4
K). This onset temperature is consistent with those
measured by resistance-temperature variations.

THERMOELECTRIC POWER

Fig. 3 shows thermoelectric power as a function of
temperature for a nominal $Tl_2Ca_2Ba_2Cu_3O_{10+X}$ sample [11].
The normal-state thermoelectric power is positive,
indicating dominant hole conduction. At least three
separate ranges of temperature-dependent behavior are
apparent. Below the transition (the midpoint of the
transition was determined to lie at 118 K), the
thermoelectric power is zero. From the transition
temperature to about 175 K, the thermoelectric power is an
increasing function of temperature. Finally, from 175 K to
room temperature, the thermoelectric power decreases
linearly with increasing temperature. The temperature
dependence of the Tl-Ca-Ba-Cu-O superconductor is
qualitatively similar to that of Y-Ba-Cu-O samples [12].

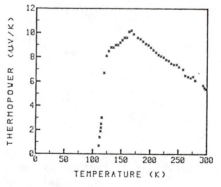

Figure 3   Thermoelectric power of a nominal
$Tl_2Ca_2Ba_2Cu_3O_{10+X}$ sample as a function of
temperature.

STRUCTURE

Two superconducting phases, $Tl_2Ca_2Ba_2Cu_3O_{10+X}$ (2223) and
$Tl_2Ca_1Ba_2Cu_2O_{8+X}$ (2122), have been isolated from the Tl-Ca-

Ba-Cu-O system [13]. The new 2223 superconductor has a 3.85 x 3.85 x 36.25 A tetragonal unit cell. The 2122 superconductor, which appears to be structurally related to $Bi_2Ca_1Sr_2Cu_2O_{8+X}$, has a 3.85 x 3.85 x 29.55 A tetragonal unit cell. The 2223 phase is related to 2122 by addition of extra calcium and copper layers. In addition, the superconducting phase in the Ca-free Tl-Ba-Cu-O system is $Tl_2Ba_2CuO_{6+X}$ (2021) [13,14]. Fig. 4 shows schematically the arrangements of metallic atoms in these three Tl-based superconducting phases. The 2021 phase has a zero-resistance temperature of about 80 K, whereas the 2122 and 2223 phases have zero-resistance temperatures 108 K and 125 K, respectively [9, 13, 14]. It appears that the addition of a Ca and Cu layer increases the transition temperature about 20 K. If this trend continues linearly, it

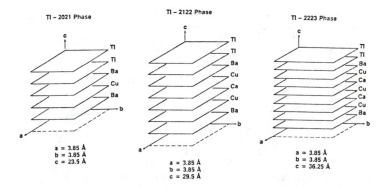

Figure 4   Schematic arrangements of the Tl-based superconducting phases 2021, 2122 and 2223.

might be expected that the 2324 phase will have a transition temperature at 140-150 K, and, based on density of states arguments within the BCS framework, a Tl-Ca-Ba-Cu-O phase with 9-10 Cu and Ca layers would have a superconducting transition above 200 K [15].

SUMMARY

The 120-K Tl-Ba-Ca-Cu-O superconductors form easily and are readily prepared. Electronic transport occurs by hole

conduction. The superconducting system consists at least of
two superconducting phases 2212 and 2223. The addition of
each Cu and Ca layer increases Tc by about 20 K. These
findings should lead to even higher temperature
superconductors.

REFERENCES

1. J.G.Bednorz and K.A.Muller, Z. Phys. B 64, 189 (1986).
2. M.K.Wu, J.R.Ashburn, C.T.Torng, P.H.Hor, R.L.Meng,
   L.Gao, Z.J.Huang, Y.Q.Wang, and C.W.Chu, Phys. Rev.
   Lett. 58, 908 (1987).
3. Z.Z.Sheng and A.M.Hermann, Nature 332, 55 (1988).
4. Z.Z.Sheng, A.M.Hermann, A.El Ali, C.Almason, J.Estrada,
   T.Datta, and R.J.Matson, Phys. Rev. Lett. 60, 937
   (1988).
5. H.Maeda, Y.Tanaka, M.Fukutomi, and T.Asano, Jpn. J.
   Appl. Phys. Lett. 27, L207 (1988).
6. C.W.Chu, J.Bechtold, L.Gao, P.H.Hor, Z.J.Huang,
   R.L.Meng, Y.Y.Sun, Y.Q.Wang, and Y.Y.Xue, Phys. Rev.
   Lett. 60, 941 (1988).
7. Z.Z.Sheng and A.M.Hermann, Nature 332, 138 (1988).
8. Z.Z.Sheng, W.Kiehl, J.Bennett, A.El Ali, D.Marsh,
   G.D.Mooney, F.Arammash, J.Smith, D.Viar, and
   A.M.Hermann, Appl. Phys. Lett. (May 16, 1988).
9. A.M.Hermann, Z.Z.Sheng, D.C.Vier, S.S.Schultz, and
   S.B.Oseroff, Phys. Rev. B (June 1, 1988).
10. S.S.P.Parkin, V.Y.Lee, E.M.Engler, A.I.Nazzal,
    T.C.Huang, G.Gorman, R.Savoyand, and R.Beyers,
    (submitted to Phys. Rev. Lett., 1988).
11. N.Mitra, J.Trefny, B.Yarar, G.Pine, Z.Z.Sheng, and
    A.M.Hermann, Phys. Rev. B. (accepted, 1988).
12. N.Mitra, J.Trefny, M.Young, and B.Yarar, Phys. Rev. B
    36, 5581 (1987).
13. R.M.Hazen, L.W.Finger, R.J.Angel, C.T.Prewitt,
    N.L.Ross, C.G.Hadidiacos, P.J.Heaney, D.R.Veblen,
    Z.Z.Sheng, A.El Ali, and A.M.Hermann, Phys. Rev. Lett.
    60, 1657 (1988).
14. C.C.Torardi, M.A.Subramanian, J.C.Calabrese,
    J.Gopalakrishnan, K.J.Morrissey, T.R.Askew,
    R.B.Flippen, U.Chowdhry, and A.M.Sleight, Science 240,
    631 (1988).
15. P.Grant (unpublished).

RELATIONSHIP OF SUPERCONDUCTING TRANSITION TEMPERATURE TO STRUCTURE IN THE Tl-Ca-Ba-Cu-O SYSTEM

S.S.P. PARKIN, V.Y. LEE and R.B. BEYERS

IBM Research Division, IBM Almaden Research Laboratory
650 Harry Road, San Jose, CA 95120-6099, U.S.A.

Abstract We have prepared a new family of high temperature superconductors of the form, $Tl_1Ca_{n-1}Ba_2Cu_nO_{2n+3}$ (n = 1, 2, 3) which contain Cu perovskite-like units separated by single TlO layers. The structures of these compounds are analogues of those previously found containing double TlO sheets. Together these six compounds provide a model family with which to examine the role of the coupling within and between the CuO_2 sheets. The transition temperatures monotonically increase with increasing number, n, of CuO_2 sheets for both the single and double TlO layer series. The transition temperatures of the TlO monolayer compounds are lower, but are similar to those for the corresponding compounds containing Bi-O bilayers, suggesting that the coupling between successive perovskite-like units is similar in these groups of materials.

INTRODUCTION

The discovery of new and more complex copper oxide superconductors has led to the recent dramatic increases in the highest known superconducting transition temperature. The original discovery of superconductivity at $\simeq 30$ K in the compound,

S.S.P. PARKIN *et al*

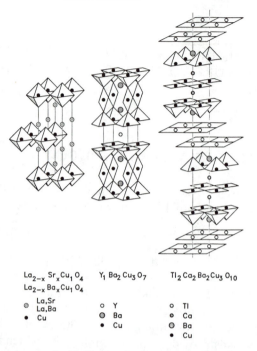

$La_{2-x}Sr_xCu_1O_4$ $Y_1Ba_2Cu_3O_7$ $Tl_2Ca_2Ba_2Cu_3O_{10}$
$La_{2-x}Ba_xCu_1O_4$

⊘	La,Sr	○	Y	○	Tl
	La,Ba	◉	Ba	•	Ca
•	Cu	•	Cu	◉	Ba
				•	Cu

Figure 1: Comparison of idealized structures of the $La_{2-x}A_xCuO_{4-y}$, $YBa_2Cu_3O_{7-x}$ and $Tl_2Ca_2Ba_2Cu_3O_{10}$ compounds.

$La_{2-x}A_x(A = Ba, Sr)CuO_{4-y}$, by Bednorz and Mueller[1] stimulated the finding of superconductivity at $T_c{\simeq}90$ K in $YBa_2Cu_3O_{7-x}$ by Wu and Chu[2] and more recently transition temperatures above 100 K in two related quinary oxide systems, Bi-Ca-Sr-Cu-O[3] and Tl-Ca-Ba-Cu-O[4-9]. All of the known high-temperature superconductors have at least one structural feature in common, namely two-dimensional CuO_2 sheets, as shown in Figure 1. The correlation of superconducting properties with distinct structural features provides an important empirical process for improved understanding of this phenomenon. In this paper[†] we discuss the properties of six related structures within the Tl-Ca-Ba-Cu-O system, one of which, $Tl_2Ca_2Ba_2Cu_3O_{10-x}$, superconducts at 125 K, at present the highest known superconducting transition temperature[6].

[†] Most of the work presented in this paper is described in more detail elsewhere[6,7,19,20].

182

Early in 1988, almost simultaneously, Maeda *et al*[3] and Sheng and Hermann[4] reported mixed-phase composites of Bi-Ca-Sr-Cu and Tl-Ca-Ba-Cu oxides respectively, with transitions to zero resistance at 85 K in the former and 107 K in the latter. The material that Maeda *et al.* prepared, of nominal composition $BiCaSrCu_2O_x$, also exhibited a diamagnetic anomaly at about 110 K. The composition of the major component, superconducting at $\simeq 85$ K, was soon identified as $Bi_2Ca_1Sr_2Cu_2O_8$ [10-12]. Several groups established that the structure comprised a copper-perovskite like unit containing two CuO_2 sheets sandwiched by Bi-O bilayers. The detailed structure of the Bi-O bilayer still remains uncertain, although it is now believed that, in contrast to some early work suggesting the structure of the Bi_2O_2 layer resembled that found in the Aurivillius phases, the oxygen and Bi atoms are approximately arranged on a rocksalt lattice[10]. It is widely believed that the higher temperature anomaly is associated with regions in the sample containing three CuO_2 sheets sandwiched by similar BiO bilayers[13-17].

Sheng and Hermann[18] originally reported superconductivity in a Tl-Ba-Cu-O composite at about 70 K, but they subsequently found that the addition of Ca increased the zero resistance temperature to just above 100 K[4]. Hazen *et al*[5] identified two distinct phases in this mixture, of nominal composition $Tl_2Ca_1Ba_2Cu_2O_8$ and $Tl_2Ca_2Ba_2Cu_3O_{10}$. By varying the preparation and starting composition of the Tl-Ca-Ba-Cu-O mixture, we[6] isolated these phases, and showed that Sheng and Hermann's original mixture was mostly comprised of the 2122 phase, which we found had $T_c \simeq 108 K$. We found that the 2223 phase had a transition to zero resistance at 125 K, with a sharp diamagnetic onset at the same temperature. These data are shown in Figures 2 and 3. The 2122 phase has an analogous structure to that of $Bi_2Ca_1Sr_2Cu_2O_8$, with the Tl and Ba replacing Bi and Sr respectively.

SAMPLE PREPARATION AND BASIC STRUCTURE

Samples of Tl-Ca-Ba-Cu-O were prepared by forming a pellet under pressure of a mixture containing appropriate proportions of Tl_2O_3, CaO, BaO_2 and CuO. The pellet was wrapped in gold and fired in a sealed quartz tube containing approximately one atmosphere of oxy-

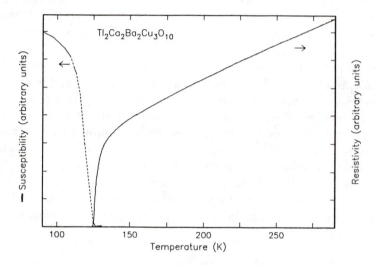

Figure 2: Plots versus temperature of resistivity and Meissner susceptibility for a pellet containing substantial amounts of the $Tl_2Ca_2Ba_2Cu_3O_{10}$ phase.

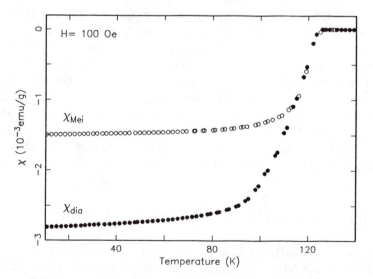

Figure 3: Meissner and diamagnetic shielding signals versus temperature for a pellet containing substantial amounts of the $Tl_2Ca_2Ba_2Cu_3O_{10}$ phase.

gen at 880°C for about 3 hours. A large variety of starting compositions was examined, and in most cases the final pellet contained several phases. However, by judicious choice of starting composition, pellets containing a single quinary phase could be prepared[6]. This then allows for a correlation of superconducting transition temperature with structure and composition. The composition and microstructure of the pellets were determined from extensive powder x-ray diffraction, electron microprobe, electron diffraction and high-resolution transmission electron microscopy (TEM) studies.

The systematic study of a large number of pellets, prepared from a wide range of starting metal cation compositions, led to the synthesis of six distinct structures of the form, $Tl_2Ba_2Cu_1O_6$ (2021), $Tl_2Ca_1Ba_2Cu_2O_8$ (2122) and $Tl_2Ca_2Ba_2Cu_3O_{10}$ (2223) and $Tl_1Ba_2Cu_1O_5$ (1021), $Tl_1Ca_1Ba_2Cu_2O_7$ (1122) and $Tl_1Ca_2Ba_2Cu_3O_9$ (1223). The idealized structures of these compounds, as determined from powder x-ray and electron diffraction and TEM, are shown in Figure 4[7,19,20]. The positions of the oxygen atoms have been inferred by comparison with reported structures of the related $La_{2-x}A_xCuO_{4-y}$, $YBa_2Cu_3O_{7-x}$ and $Bi_2CaSr_2Cu_2O_8$ compounds. The six Tl-Ca-Ba-Cu-O compounds form two distinct series, in which the copper perovskite-like units, containing one, two or three CuO_2 sheets, are separated by TlO monolayers and TlO bilayers, respectively. No analogues of the TlO monolayer compounds had previously been synthesized prior to our work on the 1223 phase[7†]. It is interesting to note the relationship of these structures to that of $YBa_2Cu_3O_{7-x}$. The role of the CuO ribbons in $YBa_2Cu_3O_{7-x}$ is taken up by the TlO mono- and bilayers. The Y^{3+} ions in $YBa_2Cu_3O_{7-x}$ are replaced by Ca^{2+} ions in the Tl-Ca-Ba-Cu-O compounds. The Ba^{2+} ions occupy similar sites in all of these structures.

The structures and compositions of these compounds are summarized in Table 1. Hereafter these compounds will be referred to by their idealized metal cation ratio given in Table 1. All six compounds have a tetragonal structure at room temperature, although the TlO monolayer compounds have a primitive tetragonal cell, whereas those

---

† During this conference we learnt that Prof. Raveau had independently synthesized the 1223 compound, which he reported was superconducting at 120 K.

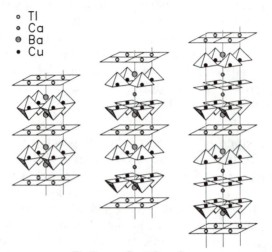

○ Tl
○ Ca
◉ Ba
● Cu

$$Tl_1 Ca_{n-1} Ba_2 Cu_n O_{2n+3}$$

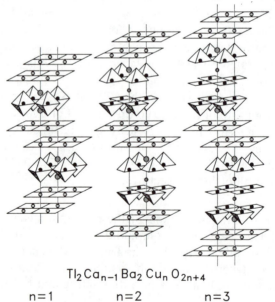

$$Tl_2 Ca_{n-1} Ba_2 Cu_n O_{2n+4}$$

$$n=1 \qquad n=2 \qquad n=3$$

Figure 4: Ideal Structures of the six $Tl_m Ca_{n-1} Ba_2 Cu_n O_{2(n+1)+m}$ phases for m = 1, 2 and n = 1, 2, 3.

Table 1: Summary of Properties of $Tl_mCa_{n-1}Ba_2Cu_nO_x$

Name (m-n)	Relative Composition Tl Ca Ba Cu O					Symmetry	Lattice Parameters a(Å) c(Å)		Superlattice wave-vector (k)	T_c(K)
$Tl_1Ca_{n-1}Ba_2Cu_nO_x$										
1021	1.2	0.0	2	0.7	4.8	P4/mmm	3.869(2)	9.694(9)	+	++
1122	1.1	0.9	2	2.1	7.1	P4/mmm	3.8505(7)	12.728(2)	<0.29,0,0.5>	65-85
1223	1.1	0.8	2	3.0	9.7	P4/mmm	3.8429(6)	15.871(3)	<0.29,0,0.5>	100-110
$Tl_2Ca_{n-1}Ba_2Cu_nO_x$										
2021	1.9	0.0	2	1.1	6.4	F/mmm¶	a = 5.445(2) b = 5.492(1)	23.172(6)	$<\overline{0.08},0.24,1>$¶	++
2021§	1.8	0	2	1.1	6.4	F/mmm¶	a = 5.4634(3) b ≃ a	23.161(1)	$<\overline{0.08},0.24,1>$¶	20
"	1.8	.02	2	1.1	6.3	I4/mmm	3.8587(4)	23.152(2)	$<\overline{0.16},0.08,1>$ =	15-20
2122	1.7	0.9	2	2.3	8.1	I4/mmm	3.857(1)	29.39(1)	<0.17,0,1>	95-108
2223	1.6	1.8	2	3.1	10.1	I4/mmm	3.822(4)	36.26(3)	<0.17,0,1>	118-125

+ No superlattice spots observed.
++ Non-metallic or weakly metallic samples with no superconducting transition observed in resistivity and magnetic susceptibility studies for temperatures down to 4.2 K.
¶ The symmetry of the structure is orthorhombic if the observed superlattice is ignored. Taking the superlattice into account lowers the symmetry to monoclinic.
= The superstructure is identical to that for the orthorhombic 2021 polymorph.
§ Sample prepared from a Cu rich starting composition, $Tl_2Ba_2Cu_2$.

containing TlO bilayers have a body-centered cell. The 2021 compound also forms a face-centered orthorhombic polymorph. The elongated unit cells of these compounds result in the low-angle portion of the powder x-ray diffraction pattern containing a single peak, which

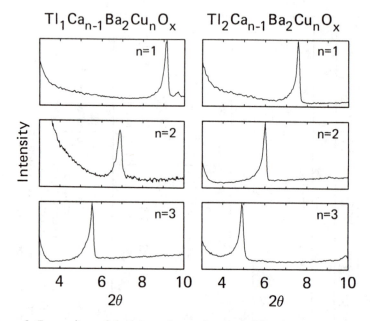

$Tl_1Ca_{n-1}Ba_2Cu_nO_x$ $Tl_2Ca_{n-1}Ba_2Cu_nO_x$

Figure 5: Comparison of the low angle powder x-ray diffraction pattern for six structures: $Tl_mCa_{n-1}Ba_2Cu_nO_{2(n+1)+m}$ $(m = 1, 2; n = 1, 2, 3)$

can be used as a useful fingerprint of the various phases. These data are compared for the six structures in Figure 5. The 001 (m = 1) or 002 (m = 2) peak systematically moves to lower Bragg angles, as the number of CuO_2 sheets is increased, and the unit cell enlarged along the c axis.

SUPERCONDUCTING PROPERTIES

The superconducting transition temperatures for the TlO monolayer and bilayer compounds are given in table 1. These temperatures were determined from Meissner susceptibility data. Typically, the magnitude of the Meissner signal corresponded to about 10-15% of the susceptibility of a perfect diamagnet of the same volume, ignoring demagnetizing corrections. In many cases the pellets contained more than one superconducting phase, and, correspondingly, the magnetic susceptibility data showed more than one transition (see, for example,

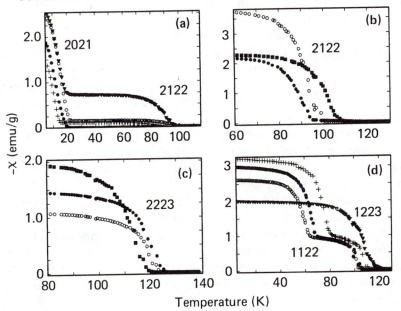

Figure 6: Meissner susceptibility versus temperature data measured in a field of 100 Oe for a number of Tl-Ca-Ba-Cu-O compounds prepared from starting compositions of (a) ($\bullet$) $Tl_2Ba_2Cu_2$, ($\circ$, $+$) $Tl_2Ca_{0.05}Ba_2Cu_{1.05}$, and ($\blacktriangledown$) $Tl_2Ca_{0.15}Ba_2Cu_{2.15}$, (b) ($\bullet$) $Tl_1Ca_2Ba_2Cu_3$, ($\circ$) $Tl_2Ca_1Ba_2Cu_2$, and ($\blacksquare$) $Tl_{2.25}Ca_1Ba_2Cu_2$, (c) ($\blacksquare$) $Tl_1Ca_2Ba_2Cu_3$, ($\circ$) $Tl_1Ca_{2.5}Ba_1Cu_3$ and ($\bullet$) $Tl_1Ca_3Ba_1Cu_3$, (d) ($\circ$) $Tl_{.85}Ca_1Ba_2Cu_2$, ($\bullet$, $+$) $Tl_1Ca_1Ba_2Cu_2$, and ($\blacktriangledown$) $Tl_{0.85}Ca_2Ba_2Cu_3$. The phases present in the pellet giving rise to the diamagnetic susceptibility are (a) 2021 and 2122, (b) 2122, (c) 2223 and (d) 1122 and 1223.

Figure 6(a) and (d)). In particular, it was difficult to prepare pellets containing only the 1122 phase. In most cases these pellets also contained some 1223 phase, and the susceptibility showed two diamagnetic transitions near 100-110 K and 65-85 K, corresponding to the 1223 and 1122 phases, respectively (Figure 6(d)). Even when the majority phase of these pellets was the 1122 phase, the resistance of these pellets would drop to zero at the higher transition temperature of the minority 1223 phase. Since, as discussed below, these materials often form intergrowths between one another, there is the potential of misassignment of transition temperatures for phases with low T_c, particularly when based solely on resistance data.

Meissner data are shown in Figure 6 for a number of representative samples. Large variations in T_c were found for material of nominally the same structure for all of the superconducting phases. For both phases containing single CuO_2 sheets, material was formed which was insulating. The 1021 phase was insulating for a wide range of preparative conditions, and no conditions were found for which it became metallic. The properties of the 2021 phase were more diverse, and varied from insulating to weakly metallic to superconducting with transition temperatures ranging from 10-20 K[†]. When prepared from a 2021 starting composition, the material had a face-centered orthorhombic unit cell and was insulating. The 2021 material became superconducting when prepared from a starting composition containing either excess Cu (Tl:Ba:Cu = 2:2:2) or small amounts of Ca and Cu in equal proportions (Tl:Ca:Ba:Cu = 2:x:2:1+x, x= 0.05-0.20). In the former case the material was weakly orthorhombic, but in the latter the material was tetragonal. The transition temperature for the Ca-doped material was independent of the Ca content in the starting composition (microprobe data revealed that the ratio of Ca/Cu in the superconducting 2021 phase was only about 0.01), but increased with the length of annealing time (at $\simeq$ 880°C).

STRUCTURAL DEFECTS

Superlattice Modulations

An important structural feature of all the Tl-Ca-Ba-Cu-O compounds, ignored above, is the presence of weak superlattice reflections in the selected area electron diffraction patterns[19-20]. Three distinct superlattice structures were found, as shown in Table 1. The reflections can be described by a set of symmetry-related wave-vectors, $\underline{k}$, where each $\underline{k}$ corresponds to a pair of reflections symmetrically disposed about each Bragg peak. The significance of the superlattice structure is unclear. There appears to be no obvious correlation with superconducting properties. For example the 1122 and 1223 compounds display the same superlattice, but the superconducting transi-

---

[†] Note that Hazen *et al*[5] and Torardi *et al*[8] report the observation of superconductivity at much higher temperatures in the 2021 phase at $\simeq$80 and $\simeq$90 K respectively.

Figure 7: Back scattered electron SEM image of a pellet of starting metal cation composition $Tl_{0.85}Ca_2Ba_2Cu_3$. The brighter regions correspond to increased Tl content compared to the neighbouring darker regions.

tion temperature of the 1223 phase is substantially higher than that of the 1122 compound. Similarly, the 2122 and 2223 phase exhibit the same superstructure, which is of the same type as that found in the 1122 and 1223 compounds. Much more intense superlattice reflections were previously observed in $Bi_2CaSr_2Cu_2O_8$, and various models have been developed to account for them[13-19]. Most models assume there is some distortion of the Bi_2O_2 sheet, with a concurrent puckering of the neighbouring CuO_2 sheets. It has been proposed that the distortion of the Bi_2O_2 layer may result from ordered vacancies, interstitial oxygen, or substitutional disorder.

Intergrowths

A common structural defect, that we have observed in all of the $Tl_mCa_{n-1}Ba_2Cu_nO_{2(n+1)+m}$ compounds, with the exception of the monolayer CuO_2 phases, is the presence of intergrowths of structures

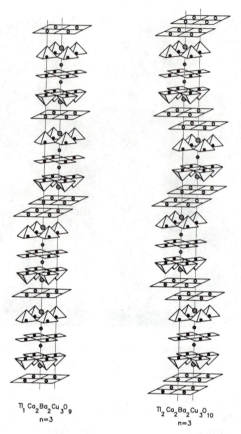

$Tl_1Ca_2Ba_2Cu_3O_9$
n=3

$Tl_2Ca_2Ba_2Cu_3O_{10}$
n=3

Figure 8: Pictorial representation of two distinct intergrowth types corresponding to (a) intergrowths of a TlO bilayer in a structure containing TlO monolayers and (b) intergrowths of double copper perovskite layers in the 2223 structure containing triple copper perovskite units.

related by the addition or deletion of CuO_2 and TlO sheets. These intergrowths are found on length scales varying from $\simeq 10{,}000$ to $\simeq 10$ Å. Figure 7 shows a back-scattered electron SEM image of crystallites of the 1223 phase in a material prepared from a metal cation starting mixture of $Tl_{0.85}Ca_2Ba_2Cu_3$. The composition of the majority of the crystallites (which comprised $\simeq 60$ % of the sample) was found by electron microprobe analysis to be $Tl_{1.05}Ca_{1.83}Ba_2Cu_{3.06}$. However, as shown in Figure 7, the SEM image of the 1223 crystallites ($\simeq$ 10-20 μ in extent) exhibits narrow bright bands, $\simeq 1\mu$ wide, consistent with

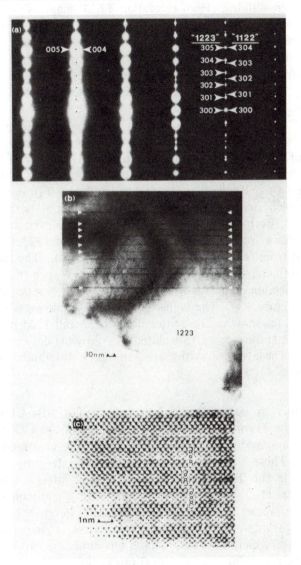

Figure 9: (a) [010] selected area diffraction (SAD) pattern and (b) corresponding image of crystallites containing regions of 1223 and 1122. The arrows in (b) denote unit cell thick intergrowths of 1122 in 1223. (c) High-resolution transmission electron micrograph of 1 unit cell thick 1122 intergrowth in 1223.

a Tl content approximately 20 % higher in these bands than in the bulk of the crystallites. High–resolution TEM images show the presence in these bands of large numbers of TlO bilayer intergrowths. A schematic representation of a way in which this intergrowth could be formed within the 1223 host lattice is shown in Figure 8. A second type of intergrowth, found in both the TlO monolayer and bilayer compounds, corresponds to one in which the number of CuO_2 sheets varies from that in the host structure. Figure 8 shows a schematic example of an intergrowth of a bilayer Cu perovskite–like unit in the 2223 structure, which contains trilayer Cu perovskite–like units. Observation of intergrowths on finer length scales is possible in the TEM, as illustrated in Figure 9 for a sample of starting composition $Tl_{0.85}Ca_1Ba_2Cu_2$. Figure 9(a) shows a selected area diffraction pattern along b^*, which clearly shows that this grain contains both 1122 and 1223 phases. Both these phases contain TlO monolayers, but whereas the 1122 phase comprises bilayer CuO_2 units, the 1223 structure is constructed from trilayer CuO_2 units (see Figure 4). The ratio of the c lattice parameters of the 1122 and 1223 phases is such, that every 4th 1122 h0l reflection coincides with every 5th 1223 h0l reflection, as can be seen in Figure 9(a). The high–resolution TEM micrographs in Figure 9(b) and (c) show intergrowths of the 1122 and 1223 phases on a finer scale extending down to isolated intergrowths only one unit cell in extent. These intergrowths are randomly distributed along the stacking axis.

Electron microprobe analysis of numerous Tl-Ca-Ba-Cu-O pellets shows that the Tl content is systematically high in the TlO monolayer compounds, and systematically low in the TlO bilayer compounds (see Table 1). These data strongly suggest that intergrowths of TlO monolayers in the TlO bilayer compounds, and intergrowths of TlO bilayers in the TlO monolayer compounds, are a common structural defect in these materials. The presence of these intergrowths, and those comprising variations in the number of CuO_2 sheets, is perhaps a natural means, by which these compounds accommodate variations in the Tl, Ca and Cu content from the ideal proportions.

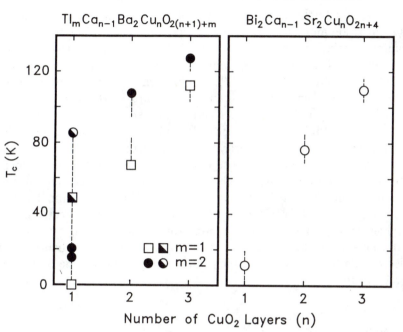

Figure 10: Dependence of T_c on the number of CuO_2 planes within the Cu perovskite-like unit for the $Tl_1Ca_{n-1}Ba_2Cu_nO_{2n+3}$ ($\square$) and $Tl_2Ca_{n-1}Ba_2Cu_nO_{2n+4}$ (● - this work, ◔ - ref 8) series of compounds. The dashed vertical lines correspond to the variations in T_c found for each phase. ◣ corresponds to data for $(Tl, Bi)_1(Ca, Sr)_2Cu_1O_x$ [21]. Data for the related $Bi_2Ca_{n-1}Sr_2Cu_nO_{2n+4}$ structures are shown for comparison[3,12,22]. The $n = 3$ $Bi_2Ca_2Ba_2Cu_3O_{10}$ phase has not yet been isolated.

RELATIONSHIP of T_c TO STRUCTURE

As observed earlier, we have found a considerable variation in T_c for all of the superconducting compounds. For example, the transition temperature of samples of nominally the 2223 phase is observed to vary from 118 to 127 K, and that of samples of the 2122 compound is found to vary from 108 to 95 K. No obvious difference in the structure of such samples, of nominally the same phase, but different transition temperatures, was found from x-ray powder diffraction or electron microprobe studies. However TEM studies did reveal that the density of intergrowths varied considerably between these samples. For the 2223 material, the highest transition temperature was found in the sample with the smallest number of intergrowths (in this material the most common intergrowth was that of bilayer CuO_2 units). A similar correlation was found for the 2122 and 1223 phases. In contrast, for the 1122 phase, the sample with the lowest density of intergrowths exhibited the lowest T_c. The extent of the variation in T_c found for these compounds is plotted in Figure 10 versus n, the number of CuO_2 layers in the copper perovskite-like unit, for both the TlO monolayer and bilayer compounds. In addition, the temperature of the sample with the lowest density of intergrowths is indicated. This temperature is presumably closest to the intrinsic transition temperature of the material. Although there appears to be a correlation between T_c and the density of intergrowths in the bilayer and trilayer Cu compounds, it is not clear that the intergrowths directly cause changes in T_c. The presence of intergrowths may simply reflect other variations in the structures, such as off-stoichiometry or substitutional disorder on the cation sites, which in turn may influence the transition temperature.

CONCLUSIONS

We have prepared a new class of high-temperature superconductors of the form, $Tl_1Ca_{n-1}Ba_2Cu_nO_{2n+3}$ (n = 1, 2, 3). The structures consist of copper perovskite-like blocks containing 1, 2, or 3 CuO_2 planes sandwiched between TlO monolayers. These compounds, plus their related structures containing TlO bilayers, thus form a model family of structures in which both the size and separation of the copper oxide blocks can be independently varied. This will allow for testing of various theoretical models of the origin of high-temperature superconductivity. This family of materials clearly establishes that the CuO_2 sheets form the essential structural ingredient for high temperature superconductivity. The superconducting transition temperature increases with the number of CuO_2 planes in the perovskite-like block for both the TlO mono- and bilayer compounds. For the same number of CuO_2 planes, the transition temperatures are consistently 15-20 K lower in the material with single TlO layers.

The predominant defects in the crystals with double and triple CuO_2 layers are stacking faults, that produce local intergrowths of related structures. The presence of stacking defects in these oxides appears to be associated with changes in the superconducting transition temperature from the values measured for defect–free samples. Stacking faults, however, do not appear to explain the variation in transition temperature in the single CuO_2 layer compound $Tl_2Ba_2Cu_1O_6$. This compound has two polymorphs, one of which is body-centered tetragonal and a superconductor, and the other is face-centered orthorhombic and an insulator. This material appears to display the widest variation in transition temperature of all the Tl-Ca-Ba-Cu-O compounds, which will provide an important focus of future studies.

ACKNOWLEDGEMENTS

We thank our colleagues, A.I. Nazzal, R. Savoy, G. Gorman, T.C. Huang and S.J. La Placa for their invaluable contributions to the work summarised in this paper. We are indebted to E.M. Engler, F. Herman and J.B. Torrance for many useful discussions.

REFERENCES

1 J.G. Bednorz and K.A. Muller, Z. Phys. B 64, 189 (1986).
2 M.K. Wu, J.R. Ashburn, C.T. Torng, P.H. Hor, R.L. Meng, L. Gao, Z.J. Huang, Y.Q. Wang and C.W. Chu, Phys. Rev. Lett. 58, 908 (1987).
3 H. Maeda, Y. Tanaka, M. Fukutomi and T. Asano, Jap. J. Appl. Phys. 27, L209 (1988).
4 Z.Z. Sheng and A.M. Hermann, Nature 332, 138 (1988).
5 R.M. Hazen, L.W. Finger, R.J. Angel, C.T. Prewitt, N.L. Ross, C.G. Hadidiacos, P.J. Heaney, D.R. Veblen, Z.Z. Sheng, A.El Ali and A.M. Hermann, Phys. Rev. Lett. 60, 1657 (1988).
6 S.S.P. Parkin, V.Y. Lee, E.M. Engler, A.I. Nazzal, T.C. Huang, G. Gorman, R. Savoy and R. Beyers, Phys. Rev. Lett. 60, 2539 (1988).
7 S.S.P. Parkin, V.Y. Lee, A.I. Nazzal, R. Savoy, R. Beyers and S.J. La Placa, Phys. Rev. Lett. (accepted).
8 C.C. Torardi, M.A. Subramanian, J.C. Calabrese, J. Gopalakrishnan, K.J. Morrissey, T.R. Askew, R.B. Flippen, U.

Chowdry and A.W. Sleight, Science 240, 631 (1988) and Phys. Rev. B (submitted).

9 M.A. Subramanian, J.C. Calabrese, C.C. Torardi, J. Gopalakrishnan, T.R. Askew, R.B. Flippen, K.J. Morrissey, U. Chowdry and A.W. Sleight, Nature 332, 420 (1988).

10 J.M. Tarascon, Y. Le Page, P. Barboux, B.G. Bagley, L.H. Greene, W.R. McKinnon, G.W. Hull, M. Giroud and D.M. Hwang, Phys. Rev. B 37, 9382 (1988).

11 S.A. Sunshine, T. Siegrist, L.F. Schneemeyer, D.W. Murphy, R.J. Cava, B. Batlogg, R.B. van Dover, R.M. Fleming, S.H. Glarum, S. Nakahara, R. Farrow, J.J. Krajewski, S.M. Zahurak, J.V. Waszczak, J.H. Marshall, P. Marsh, L.W. Rupp and W.F. Peck, Phys. Rev. B 38, July 1988.

12 J.B. Torrance, Y. Tokura, S.J. LaPlaca, T.C. Huang, R.J. Savoy and A.I. Nazzal, Solid State Comm. 66, 703 (1988).

13 T.M. Shaw, S.A. Shivanshankar, S.J. LaPlaca. J.J. Cuomo, T.R. McGuire, R.A. Roy, K.H. Kelleher and D.S. Yee, Phys. Rev. B 37, 9856 (1988).

14 D.R. Veblen, P.J. Heaney, R.J. Angel, L.W. Finger, R.M. Hazen, C.T. Prewitt, N.L. Ross, C.W. Chu, P.H. Hor and R.L. Meng, Nature 332, 334 (1988).

15 E.A. Hewatt, M. Dupuy, P. Bordet, J.J. Capponi, C. Chaillout, J.L. Hodeau and M. Marezio, Nature 333, 53 (1988).

16 H.W. Zandbergen Y.K. Huang, M.J.V. Menken, J.N. Li, K. Kadowaki, A.A. Menovsky, G. van Tendeloo and S. Amelinckx, Nature 332, 620 (1988).

17 P.L. Gai and P. Day, (preprint).

18 Z.Z. Sheng and A.M. Hermann, Nature 332, 55 (1988).

19 S.S.P. Parkin, V.Y. Lee, A.I. Nazzal, R. Savoy, T.C. Huang, G. Gorman and R. Beyers, Phys. Rev. B. (accepted).

20 R. Beyers, S.S.P. Parkin, V.Y. Lee, A.I. Nazzal, R. Savoy, G. Gorman, T.C. Huang and S.J. La Placa, Appl. Phys. Lett. (accepted).

22 C. Michel, M. Hervieu, M.M. Borel, A. Grandin, F. Deslandes, J. Provost and B. Raveau, Z. Phys. B 68, 421 (1987).

21 P. Haldar, A. Roig-Janicki, S. Sridhar and B.C. Giessen, Mat. Lett. (preprint).

THE SYNTHESIS, STRUCTURES AND PROPERTIES OF DOPED Y-Ba-Cu-M-O AND Bi-Sr-Ca-Cu-O HIGH-T$_c$ PHASES

J. M. TARASCON, P. BARBOUX, L. H. GREENE, B. G. BAGLEY, P. MICELI AND G. W. HULL
Bell Communications Research, Red Bank , NJ 07701, USA.

Abstract Homogeneous $YBa_2Cu_{3-x}M_xO_{7-y}$ (M=Fe,Co,Al,Ni and Zn) and $Bi_2Sr_2Ca_{n-1}Cu_nO_y$ (n=1,2 and 3) phases have been synthesized and characterized with respect to their structural, magnetic, and superconducting properties. We find that the oxygen content in the doped Y-based materials is dependent upon the nature and amount of doping, and reaches values greater than 7 for the Co and Fe–doped samples. Both the Fe and Co–doped systems undergo an orthorhombic-tetragonal structural transition with increased doping (x). Evidence for the site at which the metal substitution takes place is obtained by means of thermogravimetric analysis and neutron diffraction measurements. Independent of the dopant (M), T$_c$ decreases with increasing x and a correlation between the changes in T$_c$ and bond lengths is shown. The structure and properties of the superconducting bismuth phases with n=1,2 and 3 are described, and compared to those of the Y-based materials. The BiO layers in the Bi-based materials appear to play the same role (acting as a carrier reservoir and coupling the CuO$_2$ planes) as does that of the Cu-O chains in the Y-based material.

INTRODUCTION

Until recently, only a few oxides (e.g. $LiTi_2O_4$[1] and Ba-Pb-Bi-O[2]) were known

to exhibit superconductivity, and that only at relatively low temperatures

(<13K). However, the report of T$_c$'s as high as 38K by Bednorz and Muller[3]

on Ba-doped La-Cu-O, and then 90K for yttrium based cuprate oxides by Wu et

al[4] prompted much excitement among research groups throughout the world,

including Bellcore. This paper will not summarize all of our work, but instead

will focus on our studies and understanding of some crystal-chemistry aspects of

the new high-T$_c$ superconductors, with the aim to provide experimental results

that may help determine the mechanism for superconductivity, and to find

guidelines for the discovery of other high-T$_c$ materials in these oxides.

The structure of $YBa_2Cu_3O_7$ contains $Cu-O_2$ planes and Cu-O chains, with the copper denoted as Cu(2) (in planes) and Cu(1) (in chains). The oxygen located above the Cu(1) is denoted as O(4). An unanswered question has been the relative importance of the chains with respect to the planes for superconductivity in this compound. Chemical substitution at the copper sites was done in an attempt to experimentally answer this question.

For the non-oxide superconducting materials it is predicted that magnetic impurities should depress T_c faster than non-magnetic ones[5], but in our earlier work on the Zn and Ni-doped La-Sr-Cu-O (40K material) we showed that this was not the case[6]. It is important to re-examine this point with respect to the 90K materials. Reliable doping studies require homogeneous samples, especially with these new materials, for which the coherence length is only about 10Å. Because the synthesis of homogeneous samples is so important, we devote the first part of this paper to this point, and demonstrate the advantage that solution techniques have with respect to solid–state reactions. Then we discuss the 3d-metal doping at the copper sites of the $YBa_2Cu_3O_{7-x}$ phase, and show how doping affects the structural and superconducting properties.

Rare earths are not necessary to produce superconductivity in the oxide-based materials, as first indicated by Michel et al.[7], and then by Maeda et al.,[8] who reported superconductivity at 85K and a large resistance drop at 110 K in a multiphase Bi-Sr-Ca-Cu-O sample, but without identifying the superconducting phases. We investigated this system and isolated phases of general formula $Bi_2Sr_2Ca_{n-1}Cu_nO_y$ with n = 1,2 and 3, and we briefly comment here on the relationship between the structural and physical properties, and show how these properties are affected by processing conditions, namely heating time. We also compare our results to those of the 123 system, to emphasize that the presence of Cu-O chains is not required to obtain a high T_c.

SYNTHESIS

The synthesis of a $YBa_2Cu_3O_7$ superconducting ceramic consists of mixing stoichiometric amounts of Y_2O_3, CuO and either barium carbonate or nitrate, and firing the mixture at temperatures ranging from 920 to 970°C. The T_c's of the resulting material depend strongly upon both the processing atmosphere and the cooling rate, with the highest T_c's achieved in slowly cooled samples and an oxygen ambient. The resulting solid–state reaction–derived ceramics have low densities (50-60%) and have T_c's of 92K with transition widths of 1K or more. They also exhibit very low critical current densities (J_c) of about $200A/cm^2$ at 77K, in contrast to the values of 10^5 obtained on thin films. Sample inhomogeneity is one of the probable causes for the low J_c's. Thus, in an attempt to overcome this problem, we have investigated possible solution techniques, and have developed a sol-gel process[9-10], in which the components are mixed at the atomic level.

The gelation procedure for the 123 phase involves the mixing of colloidal $Y(OH)_3$ with appropriate amounts of copper acetate, barium hydroxide and barium acetate so that the solution pH is close to 7. After a few hours, blue viscous gels form, which are stable for several weeks. The gels can be dried at room temperature to produce a glassy-like material which, after being heated to 920°C, produces a powdered 123 material having a small particle size (1 μm). Using this technique, the synthesis temperature of the 123 material can be lowered to 840°C. At temperatures lower than this, the samples are always contaminated by a trace amount of barium carbonate (as determined by x-ray diffraction). These powders can be pressed into pellets and re-sintered to produce a bulk dense ceramic. Using such a process with inexpensive, reagent grade materials, we are able to prepare a 90% dense ceramic, whose superconducting properties are shown in Figure 1. Note that T_c is about 92K and the transition width is only 0.6K. Even though the transition is sharper, an indication of homogeneity, the critical current remains low, never exceeding 10^3 A/cm^2 at 77K. We believe that carbon, which can be present in both solid-state and solution reactions, leads to the formation of stable $BaCO_3$ prior to the formation of the 123 phase, and this carbonate (at grain boundaries) reduces, and limits, the critical current. The $YBa_2Cu_{3-x}M_xO_{7-y}$ (M = Fe, Ni, Co, Al and Zn) materials discussed next were prepared using both the solid-state and solution techniques described above.

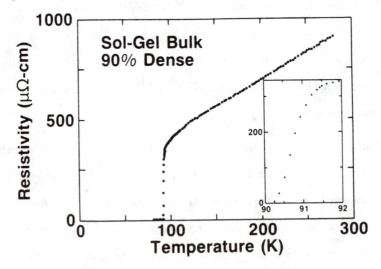

FIGURE 1. The resistivity vs. temperature is shown for a 90% dense ceramic derived using a solution technique. (ref. 10)

DOPED Y–Ba–Cu–M–O AND Bi–Sr–Ca–Cu–O HIGH–T_c PHASES

STRUCTURAL AND PHYSICAL PROPERTIES OF $YBa_2Cu_{3-x}M_xO_{7-y}$ (M = Fe, Co, Al, Ni and Zn)

Structural studies have shown (Fig. 2) that Fe, Al and Co ions induce an orthorhombic-tetragonal transition (i. e. "a" become equal to "b") within the range of composition $(0.5 < x < 0.1)$, whereas the unit cell remains orthorhombic for M = Ni or Zn over the entire range of solubility. Aluminum ions are unequivocally trivalent. Therefore, the similar structural behavior observed for the Al, Co and Fe series strongly suggests that both Fe and Co ions are also trivalent. The trivalent Fe, Co and Al ions are smaller than divalent or trivalent Cu, so one would expect a decrease of the cell volume with increasing dopant content. This is contradicted by experimental data, which shows an increase in "c" with increasing x, suggesting that the 3d-metal doping may change the stoichiometry of another constituent. In previous work[12] we reported the inability to achieve oxygen contents greater than 7 with undoped 90K material (even at low temperatures). However, chemical analysis and TGA measurements, performed on the 3d-metal doped materials, have shown oxygen contents greater than 7, for values of x greater than 0.4, for both the Fe and Co doped series.

It is well established that undoped material can reversibly accommodate one oxygen atom per formula unit, and that this mobile oxygen belongs to the Cu-O chains[13-14]. This material can thus be viewed as an intercalation compound, and any chemical perturbation in the Cu-O chains can be expected to affect the removal of oxygen, whereas no effect is expected for chemical substitutions taking place in the $Cu\text{-}O_2$ planes. A study of oxygen removal in this series of

203

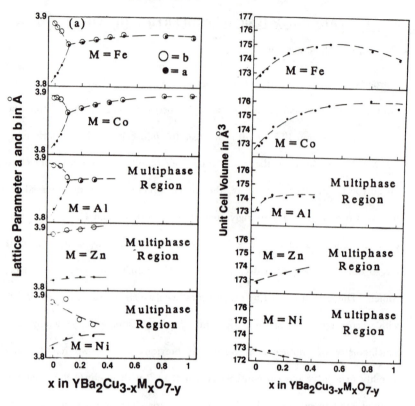

FIGURE 2. The unit cell parameters are reported for the $YBa_2Cu_{3-x}M_xO_{7-y}$ series with M = Fe, Co, Al, Ni and Zn. (a). "a" and "b" vs. x are shown. (b). the unit cell volume (V) vs. x is shown.

compounds by means of thermogravimetric analysis measurements is shown in Figure 3. The amount of oxygen removed from the compound decreases with increasing x for the Co and Fe series, whereas it remains constant for the Ni and Zn substitutions. Note also that, for the Fe series, the removal of oxygen is constant for x greater than 0.5. We thus conclude that; (1) Co substitutes only on the Cu(1) sites; (2) at low concentration Fe goes preferentially on the Cu(1) sites, whereas at higher concentrations (x > 0.5) some occupation occurs on the

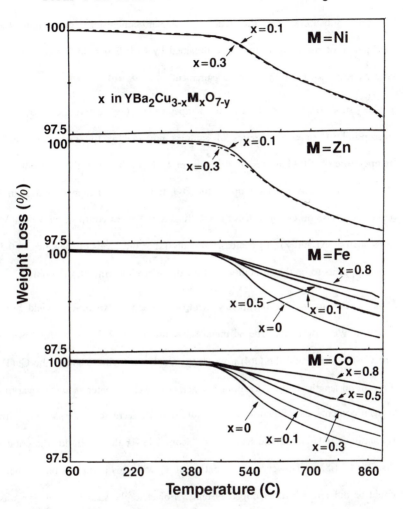

FIGURE 3. The TGA traces are shown as a function of x for the YBa$_2$Cu$_{3-x}$M$_x$O$_{7-y}$ (M = Ni, Zn, Fe and Co). The samples were heated in an argon atmosphere up to 900°C at a rate of 10°C/min. [Ref. 11]

Cu(2) sites; (3) for both Ni and Zn (independent of the concentration) the substitution seems to occur on the Cu-O$_2$ planes, but no definitive conclusion can be made. Identical experiments done on the Al series samples showed that

Al only substitutes for the $Cu(1)$[11]. Similar conclusions concerning the occupancy of the copper sites were obtained by Mössbauer (in the case of Fe) and Neutron powder diffraction measurements[11-15].

The superconducting transition temperatures (as determined by ac-susceptibility) for several series of samples prepared at the same time at a temperature of 970°C are reported in Figure 4. Note that, with the trivalent ions, T_c remains roughly constant up to the O-T transition, and then decreases in a continuous fashion down to less than 4.2K at x = 0.5. In contrast, with divalent ions such as Ni and Zn, T_c decreases initially faster, but levels off beyond x = 0.3, which appears to be the solubility limit for the annealing temperature used.

In addition to the site occupancy, neutron diffraction experiments yield bond lengths. We observe that the 3d metal substitution moves the Ba atoms towards the $Cu-O_2$ planes and the O(4) towards the $Cu(1)$[15]. Figure 5 shows the Cu(1)-O(4) bond length changes, for doped materials as well as materials having various oxygen contents, as a function of T_c. Note that T_c decreases systematically with decreasing Cu(1)-O(4) bond length, and becomes lower than 4.2K for the critical distance of 1.82Å. However, the variation of T_c with this particular bond length could be different for other doped systems. For example, such a correlation was not found in the $NdBa_{2-x}Nd_xCu_3O_{7-y}$ series[16].

There has been much experimental work dealing with the effects of a 3d metal substitution, and some results differ from group to group[17-18]. To understand the origin of such a discrepancy, we prepared another series of Ni and Co-doped samples at lower temperatures. The T_c's of these samples are shown in Figure 6

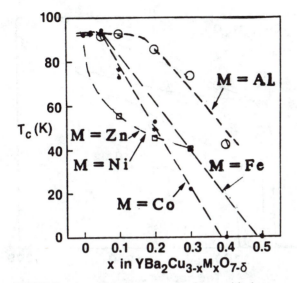

FIGURE 4. The superconducting critical temperatures, determined inductively, are reported as a function of x for Fe, Co, Ni, Zn and Al-doped 90K perovskite series. [Ref. 11]

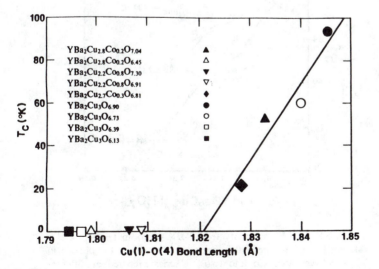

FIGURE 5. The T_c's are plotted as a function of the Cu(1)-O(4) distance for 3d-metal doped sample and undoped oxygen-deficient samples. [Ref. 15]

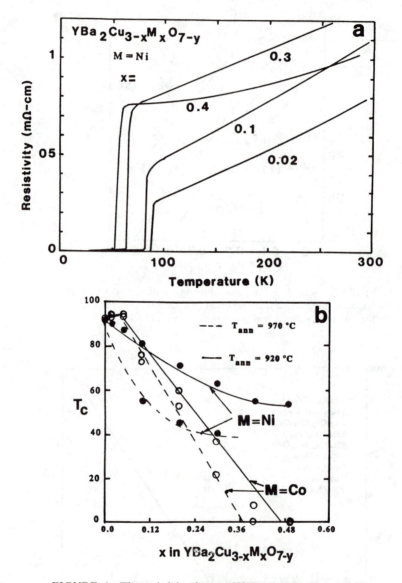

FIGURE 6. The resistivity from 4.2K to room temperature is shown for a Ni-doped 90K series (a). (b) shows the change in T_c's between two Co and two Ni series prepared at different temperatures.

together with the resistivity curves for several compositions. We observed that, as the annealing temperature is lowered, the apparent solubility range is extended to higher x, and T_c decreases less sharply. What is correlated is the nominal composition vs. T_c; we cannot preclude the possibility that minor second–phase precipitates rich in the dopant drive the major phase composition to smaller x. However, our experimental observations are consistent with our inability to substitute Ga for Cu at 970°C, whereas others[17] were able to make the substitution at a lower temperature. We always observe a sharper decrease for the divalent than the trivalent ions, at least at low x. The superconducting transitions, determined resistively for the Ni–doped samples, are relatively sharp ($\Delta T = 3K$), but more interesting is the change in the metallic behavior of the resistivity above T_c. As y increases from 0.02 to 0.4, both T_c and $d\rho/dT$ are observed to decrease. A similar behavior was found with the Sr–doped La_2CuO_4 materials[19].

Thus, the superconducting properties of the doped samples are extremely dependent upon thermal history. However, independent of the heating temperature, a general observation is that, at low x, T_c always decreases faster with concentration when the doping ions are divalent. Substitution of the divalent ions at the Cu(2) sites, as indicated by our data, is consistent with superconductivity being associated with the planes. The chains can then be viewed as a reservoir of holes, assuming a charge transfer between the $Cu-O_2$ planes. Either the removal of oxygen, or a 3d-metal substitution, results in a simultaneous motion of the O(4) towards the chains and the Ba towards the plane, such that the $Cu-O_2$ planes become isolated (i.e no charge transfer). As a result the material becomes a semiconductor, and antiferromagnetism appears (due to electronic localization). Thus a chemical unit (not necessarily a Cu-O chain) seems to be necessary to act as a hole reservoir to provide the electronic coupling between the $Cu-O_2$ planes in these new cuprate oxides. This point is even better illustrated with the Bi-based materials where T_c's as high as 110K are observed in a structure exempt of Cu-O chains, as is now described.

THE Bi-Sr-Ca-Cu-O SYSTEM

We observed that the $Bi_4Sr_3Ca_3Cu_4O_y$ phase (denoted 4334) was responsible for superconductivity at 85K in this system[20]. Single–phase 4334 compounds were made by firing in air a stoichiometric mixture of bismuth oxide, copper oxide and the corresponding Sr and Ca carbonates within the temperature range 850-860°C. Crystals were obtained using excess Bi and Cu oxides as a flux. Small crystals were used to determine the crystal structure of this phase. The pseudotetragonal substructure can be viewed as a three-dimensional packing of $Bi_2Sr_2CaCu_2O_8$ slabs (which consists of a Ca layer surrounded by 2 CuO_2, 2SrO and 2 BiO layers) crystallographically sheared with respect to each other. The exact structure is more complicated, because of an incommensurate modulation in the 010 direction. The 2-dimensional, micaceous character of this material led us to expect, as commonly observed in other layered–type materials, that several phases, having different stacking sequences, might exist, and might be stabilized by changing either the annealing conditions or the nominal composition.

We showed, for instance, that heating the 4334 phase close to its melting point produces a decomposition (described elsewhere in more detail)[20,21] which results in a multiphase sample containing the 110K phase of composition 2223 as the major phase, and for which crystals can be isolated. The x-ray diffraction pattern of a highly c-oriented crystal composite is shown in Figure 7. Only the 001 Bragg peaks are observed. In comparison to the 4334 phase, note that the 002 Bragg peak is shifted to a lower angle, thus indicating a larger c-axis, 37 instead of 30.6Å.

We find that heating samples of nominal composition 2201 in air at temperatures of 860-890°C produces a multiphase sample, with the major phase having a low angle Bragg peak located at $2\Theta = 7.15°$, corresponding to the 2201 phase. We were not able to eliminate the minor phase if preparation was in an open container, but we found that the major phase becomes unique when a mixture of Bi_2O_3,SrO_2 and CuO powders (in the 2201 ratio), placed in an alumina boat, was fired at 840-850°C for 15 hours in a sealed quartz ampoule. The resulting samples were semimetallic and not superconductors, but superconductivity could be induced by operating at higher temperatures (860-870°C), whereupon traces of the minor phase again appear in the x-ray powder pattern. Mixtures of nominal composition 2201 were melted at T = 1140°C, cooled to 890°C and maintained at this temperature for several days. Platelet-like crystals, corresponding to the 2201 phase, were sticking out of the bulk. These crystals, whose x-ray diffraction pattern is shown in Figure 7, were superconductors at 10K. We also note that, using this preparation technique, we found evidence for other related phases, with Bragg peaks at even lower angles (2 $\Theta = 6°$ and 4.9° using CuKα), but we were not able to isolate them. Resistivity measurements on bulk compacts containing these extra phases indicates a semiconducting-type behavior down to 4.2K.

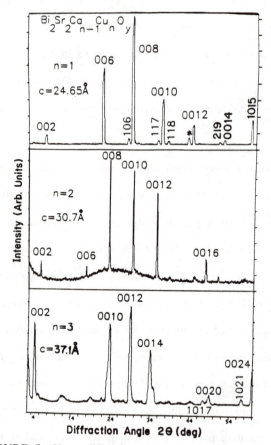

FIGURE 7. X-ray diffraction patterns collected on highly c-oriented composites of the 2201, 2212, and 2223 phases. Only the 001 lines are present.

The structures of the $Bi_2Sr_2Ca_{n-1}Cu_nO_y$ phases with n = 1,2 and 3 are shown in Figure 8. All of these phases crystallize in a pseudotetragonal unit with a = 3.83 $\sqrt{2}$ and c increasing from 24.6 to 30.7 and 37.1Å in going from n = 1 to 2 and 3. These phases differ only by the number of CuO_2-Ca-CuO_2 units within the repeat slab (c/2). The increase in the c-axis observed in going from the n = 1 to n = 3 phase corresponds to the addition of 2 times 1 and 2 $CaCu_2O_4$ units, each about 3Å thick.

212

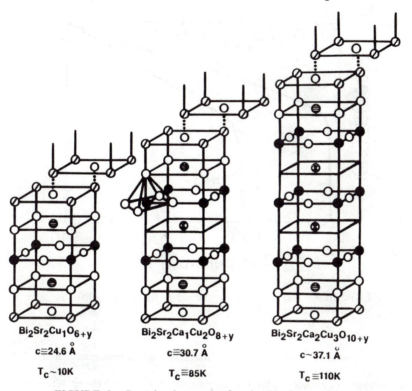

Bi$_2$Sr$_2$Cu$_1$O$_{6+y}$

c≡24.6 Å

T$_c$~10K

Bi$_2$Sr$_2$Ca$_1$Cu$_2$O$_{8+y}$

c≡30.7 Å

T$_c$≡85K

Bi$_2$Sr$_2$Ca$_2$Cu$_3$O$_{10+y}$

c~37.1 Å

T$_c$≡110K

FIGURE 8. Crystal substructure for the Bi phases of general formula Bi$_2$Sr$_2$Ca$_{n-1}$Cu$_n$O$_y$ with n = 1, 2 and 3.

Resistivity measurements on crystals of the three Bi phases are shown in Figure 10. T$_c$ increases from 10K (n = 1) to 85K (n = 2) and 110K for n = 3, and therefore there is not at a linear dependence between T$_c$ and the number of Cu-O$_2$ layers. Although we did not address it here, there can be large changes in T$_c$ (of about 10K) depending upon the thermal process used. The stacking faults detected by TEM measurements in most of the samples are likely responsible for such changes, as discussed elsewhere[22].

CONCLUSIONS

Chemical doping studies of the Y-based compound has shown both that T$_c$ is

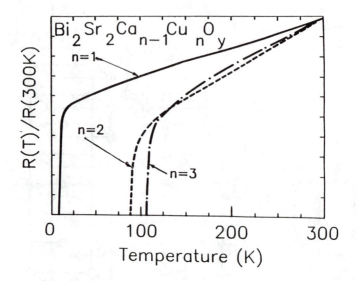

FIGURE 9. The resistivity normalized to the value at 300K is shown for the three Bi-based phases.

depressed whether the dopant is magnetic or nonmagnetic, and that superconductivity resides in the $Cu-O_2$ planes, with the Cu-O chains acting only as a reservoir of holes, with a charge transfer to the planes. In the Bi-based phases the charge reservoir is the BiO layer. However, with this system there is less freedom to modify the carrier concentration. The new Bi phases have a layered structure. Thus, as with other layered compounds, we find that processing time, temperature, and cooling rate are critical parameters in the preparation of the n = 1,2 and 3 phases.

ACKNOWLEDGEMENTS

We thank P. A. Morris, W. R. McKinnon, Y. LePage, J. M. Rowell and J. H. Wernick for helpful discussions.

REFERENCES

1. D. C. Johnson, H, Prakash, W. H. Zachariasen and R. Viswanathan, Mat Res. Bull 8, 777 (1973).

2. A. W. Sleight, J. L. Gillson, F. E. Biersled, Solid State Commun. 17, 27 (1975).

3. J. G. Bednorz and K. A. Müller, Z. Phys. B64, 189 (1986).

4. M. K. Wu, J. R. Ashburn, C. J. Torng, P. H. Hor, R. L. Meng, L. Gao, Z. J. Huang, Y. Q. Wang, and C. W. Chu, Phys. Rev. Lett. 58fR, 908 (1987).

5. A. A. Arbrikosov and L. P. Gork'ov, Zh. Eksp. Teor. Fiz., 29, 1781 (1960); Sov. Phys. JETP (Engl. Transl.), 12, 1243 (1961).

6. J. M. Tarascon, L. H. Greene, P. Barboux, W. R. McKinnon, G. W. Hull, T. P. Orlando, K. A. Delin, S. Foner and E. J. McNiff, Phys. Rev. B 36,8393 (1987).

7. C. Michel, M. Hervieu, M. M. Borel, A. Grandin, F. Deslandes, J. Provost and B. Raveau, Z. Phys. B68, 421 (1987).

8. H. Maeda, Y. Tanaka, M. Fukutomi and T. Asano, Jpn. J. Appl. Phys. Lett. 27, L209 (1988).

9. P. Barboux, J. M. Tarascon, L. H. Greene, G. W. Hull and B. G. Bagley, J. Appl. Phys., 63, 2725 (1988).

10. P. Barboux, J. M. Tarascon, B. G. Bagley, L. H. Greene, G. W. Hull, B. W. Meagher and C. B. Eom, Mat. Res. Symp. Proc. 99, 49 (1988).

11. J. M. Tarascon, P. Barboux, P. Miceli, L. H. Greene and G. W. Hull Phys. Rev. B, 37, 7458 (1988).

12. B. G. Bagley, L. H. Greene, J. M. Tarascon and G. W. Hull, Applied Phys. Lett. 51, 622 (1987).

13. J. M. Tarascon, W. R. McKinnon, L. H. Greene, G. W. Hull, and E. M. Vogel, Phys. Rev. B36, 16 (1987).

14. J. M. Tarascon, P. Barboux, B. G. Bagley, L. H. Greene, W. R. McKinnon and G. W. Hull, in **Chemistry of High Temperature Superconductors**, D. L. Nelson, M. S. Whittingham and T. F. George, eds., (Amer. Chem. Soc., Washington, D. C. 1987). p198

15. P. F. Miceli, J. M. Tarascon, L. H. Greene, P. Barboux, F. J. Rotella and J. D. Jorgenson, Phys. Rev. B37, 5932 (1988).

16. D. G. Hinks, Bull. Am. Phys. Soc. 33, 610 (1988).

17. G. Xiao, F. H. Streitz, A. Gavrin, Y. W. Du, and C. L. Chien, Phys. Rev. B35, 8782 (1987).

18. Y. Maeno, T. Tomita, M. Kyogoku, S. Awaji, Y. A. Oki, K. Hoshino, A. A. Minami and T. Fujita, Nature 328, 512 (1987); and Y. Maeno, M. Kato, Y. Aoki and T. Fujita, Jpn. J. Appl. Phys., 266, L1982 (1987).

19. J. M. Tarascon, L. H. Greene, W. R. McKinnon, G. W. Hull and T. H. Geballe, Science 235, 1373 (1987).

20. J. M. Tarascon, Y. LePage, P. Barboux, B. G. Bagley, L. H. Greene, W. R. McKinnon, G. W. Hull, M. Giroud and D. M. Hwang, Phys. Rev. B37, 9382 (1988).

21. J. M. Tarascon, Y. LePage, L. H. Greene, B. G. Bagley, P. Barboux, D. M. Hwang, G. W. Hull, W. R. McKinnon and M. Giroud, Phys. Rev. B July 1, 1988. (in press)

22. J. M. Tarascon, W. R. McKinnon, P. Barboux, D. M. Hwang, B. G. Bagley, L. H. Greene, G. W. Hull, Y. LePage, N. Stoffel and M. Giroud, Phys. Rev. B (submitted).

SYNTHESIS, STRUCTURE, AND PROPERTIES OF $A_2B_2Ca_{n-1}Cu_nO_{2n+4}$ SUPERCONDUCTORS (A/B = Bi/Sr or Tl/Ba, and n = 1, 2, 3)

C. C. TORARDI, M. A. SUBRAMANIAN,
J. GOPALAKRISHNAN, E. M. McCARRON,
J. C. CALABRESE, K. J. MORRISSEY, T. R. ASKEW,
R. B. FLIPPEN, U. CHOWDHRY AND A. W. SLEIGHT
Central Research and Development Department
E. I. du Pont de Nemours and Company
Experimental Station
Wilmington, Delaware 19898, U.S.A.

and

D. E. COX
Physics Department
Brookhaven National Laboratory
Upton, New York 11973, U.S.A.

Abstract   There is now a new series of high temperature superconducting oxides that may be represented as $A_2B_2Ca_{n-1}Cu_nO_{2n+4}$ where A/B = Bi/Sr or Tl/Ba, and n, the number of CuO_2 sheets stacked consecutively in the structure, equals 1, 2, or 3. There is a general trend toward higher superconducting transition temperatures as n increases. The preparative conditions, superconducting properties, and structures of these compounds are discussed.

INTRODUCTION

High-temperature superconductivity has gained renewed attention following recent reports of onset temperatures above 100 K in the new class of Bi-Sr-Ca-Cu-O[1,2] and Tl-Ba-Ca-Cu-O[3] ceramics. Superconducting transition temperatures, T_c, of up to ~125 K are found in these materials. The manner

217

in which the samples are prepared as well as their structural properties are important in determining T_c. It is now apparent that there is a series of superconducting copper-based oxides that may be represented as $A_2^{III} B_2^{II} Ca_{n-1}Cu_nO_{2n+4}$ where A^{III} is Bi or Tl, B^{II} is Sr (when A = Bi) or Ba (when A = Tl), and n (= 1, 2, or 3) is the number of consecutive Cu-O layers[4]. The structures are comprised of single, double, or triple CuO_2 sheets separated by calcium ions (for the n = 2 and 3 compounds). In all cases, the copper-oxide units stack with sheets of strontium or barium ions, and double sheets of Bi-O or Tl-O atoms. Relationships between structure and superconducting properties have been investigated in detail and are summarized here.

EXPERIMENTAL

Synthesis

Powder samples in the Bi-Sr-Ca-Cu-O[5,6] and Tl-Ba-Ca-Cu-O[4,5,7] systems were prepared by heating at 850-925°C for 1-36 hours stoichiometric mixtures of high-purity oxides in open crucibles for the Bi/Sr compositions, and in sealed gold tubes for the Tl-Ba compositions. The preparative conditions are summarized in Table I. Crystals were synthesized by heating off-stoichiometric (copper-rich) mixtures as shown in Table I. Superconducting transition temperatures were determined by flux exclusion and four-probe electrical resistivity measurements. In general,

TABLE I Summary of the preparative conditions for $A_2B_2Ca_{n-1}Cu_nO_{2n+4}$ compounds (A/B = Bi/Sr or Tl/Ba, and n = 1, 2, 3)

Compound	Reactants/Stoichiometry	Conditions	$T_c(K)$[a]
$Bi_2Sr_2CuO_6$	Bi_2O_3, SrO_2, CuO	gold crucible	
powder	1:2:1	850°C, 12 hrs	n.c.[b]
crystals	1:2:3	950°C, 12 hrs	12
$Bi_2Sr_2CaCu_2O_8$	Bi_2O_3, SrO_2, $CaCO_3$, CuO	gold crucible	
powder	1:2:1:2	850°C, 12–48 hrs	85
crystals	1:2:1:3	875°C, 36 hrs	95
$Tl_2Ba_2CuO_6$	Tl_2O_3, BaO_2, CuO	sealed gold tube	
powder	1:2:1	875°C, 3 hrs	84
crystals	1:2:2	900°C, 9 hrs	90
$Tl_2Ba_2CaCu_2O_8$	Tl_2O_3, BaO_2, CaO_2, CuO	sealed gold tube	
powder	1:2:1:2	900°C, 6 hrs	98
crystals	1:2:1:3	915°C, 15 min	110
$Tl_2Ba_2Ca_2Cu_3O_{10}$	Tl_2O_3, BaO_2, CaO_2, CuO	sealed gold tube	
powder	1:2:2:3	890°C, 1 hr	105
crystals	1:2:6:6	920°C, 3 hrs	125

[a] Onset temperature from flux exclusion measurements.
[b] not superconducting

it was found that partially melted or melted samples give higher T_c's. However, x-ray powder diffraction patterns of these samples often showed a mixture of phases. This is discussed in further detail below.

Crystallography

Crystals were examined by electron microscopy as well as x-ray diffraction, using both precession photographs and axial oscillation photographs. Single crystal x-ray diffraction data were obtained as previously described for $Bi_2Sr_2CuO_6$[5], $Bi_2Sr_2CaCu_2O_8$[6], $Tl_2Ba_2CuO_6$[5], $Tl_2Ba_2CaCu_2O_8$[7], and $Tl_2Ba_2Ca_2Cu_3O_{10}$[4]. Neutron powder diffraction data for the n = 2 and 3 Tl/Ba phases were obtained at Brookhaven National Laboratory. Unit cell parameters for these phases are given in Table II.

The $Bi_2Sr_2Ca_{n-1}Cu_nO_{2n+4}$ (n = 1, 2) crystals exhibit a micaceous morphology and are easily cleaved[5,6]. They possess orthorhombic subcell symmetry, and display satellite reflections along one of the basal axes (the a-axis) indicating the existence of a superstructure with a dimension of ~ 5a. The supercell can appear to be along both the a and b axes, due to 90° misorientations. A superstructure along the c axis of the n = 1 material, with a periodicity of approximately 3 times the c dimension, is also observed. It was not possible to determine the precise atomic arrangement of the double Bi-O layers, and it is assumed that the supercells arise from an ordering of these Bi and O atoms.

220

SYNTHESIS OF $A_2B_2Ca_{n-1}Cu_nO_{2n+4}$ SUPERCONDUCTORS

In comparison, crystals of $Tl_2Ba_2Ca_{n-1}Cu_nO_{2n+4}$ (n = 1, 2, 3) are plate-shaped but not micaceous[4,5,7]. No obvious superstructure is observed, and no twinning occurs because of the tetragonal symmetry. The Tl-O double layers are closer to an ideal structure than the analogous layers in the Bi/Sr phases, but a local disorder is suggested from the structural refinement (discussed below). Lattice imaging of the n = 3 phase clearly shows the Tl-O double layers and the Cu-O triple layers[4]. However, a prominent defect in some of the particles was the presence of five consecutive Cu-O layers. Electron diffraction from regions of intergrowth show a c-axis spacing of 48 Å, the expected value for an n = 5 phase.

TABLE II   Crystallographic information for $A_2B_2Ca_{n-1}Cu_nO_{2n+4}$ (A/B = Bi/Sr or Tl/Ba, and n = 1, 2, 3) compounds

Compound[a]	a(Å)	b(Å)	c(Å)
$Bi_2Sr_2CuO_6$	5.362(3)	5.374(1)	24.622(6)
$Bi_2Sr_2CaCu_2O_8$	5.399(2)	5.414(1)	30.904(16)
$Tl_2Ba_2CuO_6$	3.866(1)		23.239(6)
$Tl_2Ba_2CaCu_2O_8$	3.8550(6)[b] 3.8559(1)		29.318(4) 29.420(1)[b]
$Tl_2Ba_2Ca_2Cu_3O_{10}$	3.8503(6)[b] 3.8487(1)		35.88(3) 35.662(2)[b]

[a]Space group Amaa for Bi/Sr compounds, I4/mmm for Tl/Ba compounds.
[b]From neutron powder diffraction data, all others from single crystal x-ray data.

Structures Of The Bi/Sr Compounds

The structure of $Bi_2Sr_2CuO_6$[5] contains single

221

sheets of corner-sharing CuO_4 units oriented in the (001) plane, in which each copper atom has two additional oxygen atoms positioned above and below the sheet to form an axially elongated (Jahn-Teller-distorted) octahedron (Figure 1). As found in La_2CuO_4, the octahedra are alternately tilted within the copper-oxygen sheets. This tilting of the CuO_6 units appears to be responsible for the distortion from tetragonal to orthorhombic symmetry in La_2CuO_4[8] and may contribute to the observed orthorhombic symmetry in $Bi_2Sr_2CuO_6$.

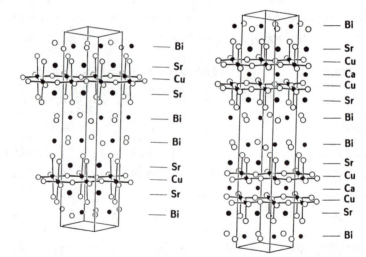

$Bi_2Sr_2CuO_6$ $Bi_2Sr_2CaCu_2O_8$

FIGURE 1. Structure of $Bi_2Sr_2CuO_6$ and $Bi_2Sr_2CaCu_2O_8$. Metal atoms are shaded and Cu-O bonds are shown.

$Bi_2Sr_2CaCu_2O_8$[6] possesses double sheets of corner-sharing CuO_4 units (Figure 1). The Cu-Cu separation between the sheets is 3.25 Å. There are additional oxygen atoms located above and below the double sheets, creating a square pyramidal environment around each copper atom. Calcium (and some strontium or bismuth) ions are located between the Cu-O sheets in eight-fold coordination with oxygen. This structural feature is also found in the $RBa_2Cu_3O_7$ phases where the copper-oxygen sheets are separated by 3.3 Å with trivalent rare earth cations[9].

Strontium cations reside just above and below the single and double Cu-O sheets of the 2210 and 2212 compounds, respectively. These copper-oxygen/alkaline-earth slabs alternate with a double bismuth-oxygen layer, but, because of apparent disorder in the subcells of these materials, the precise nature of the Bi-O bonding is not yet understood.

Chemistry of the Bi/Sr Materials

Flux exclusion measurements for stoichiometric $Bi_2Sr_2CaCu_2O_8$ powder shows bulk superconductivity at roughly 85 K. However, in similar experiments, superconducting onset temperatures of ~117 K have been noted[6]. In light of recent results in the $Tl_2Ba_2Ca_{n-1}Cu_nO_{2n+4}$ system, for which T_c increases with increasing n, the high onset temperature is most likely related to the formation of regions of $Bi_2Sr_2Ca_2Cu_3O_{10}$ (n = 3) within these predominantly $Bi_2Sr_2CaCu_2O_8$ samples.

We have observed that $Bi_2Sr_2CaCu_2O_8$ decomposes with increasing temperature according to:

223

$$Bi_2Sr_2CaCu_2O_8 \longleftrightarrow Bi_2Sr_2CuO_6 + [CaCuO_2] \quad (1)$$

It has been demonstrated that the addition of excess CaO and CuO drives the above reaction to the left and hence promotes the formation of the 2212 phase. Similarly, further addition of CaO and CuO might possibly stabilize even higher n values of $Bi_2Sr_2Ca_{n-1}Cu_nO_{2n+4}$. When the nominal starting compositional n value is increased, the bulk transition temperature remains constant at ~85 K, while a portion of the sample, which is probably n = 3 material, superconducts at ~117 K. The fact that the bulk of the material remains superconducting at 85 K (n = 2) implies that there is a delicate balance between the thermodynamic and kinetic factors at play in the layered Bi/Sr system:

$$Bi_2Sr_2Ca_2Cu_3O_{10} \longleftrightarrow$$
$$Bi_2Sr_2CaCu_2O_8 + [CaCuO_2] \longleftrightarrow$$
$$Bi_2Sr_2CuO_6 + 2[CaCuO_2] \quad (2)$$

In short, increasing the temperature drives the lefthand equations to the right, while increasing the CaO and CuO concentrations drives the righthand equations to the left. One implication of this is that low temperature ought to favor the formation of high n phases, but unfortunately the temperatures required for solid state synthesis are already too high to form pure n = 3 (117 K) material. Apparently, novel low temperature synthesis routes must be discovered in order to

SYNTHESIS OF $A_2B_2Ca_{n-1}Cu_nO_{2n+4}$ SUPERCONDUCTORS
push T_c higher by increasing n.

Another way to increase the T_c of the bismuth phases is through off-stoichiometric melting. In all cases, melting leads to the formation of complex multiphase mixtures. For the melted material containing excess CuO and CaO, there is no evidence of the 117 K 2223 phase – consistent with the high temperature driving equation 2 to the right, and resulting in the formation of a 2212-like phase. However, the T_c of this material is found to be ~95 K, rather than the ~85 K typically associated with 2212 materials made at lower temperatures. The cause of this increased T_c is as yet unknown, but, by analogy with the following work on lead substitution, it is likely due to cation substitutions.

Incorporation of lead on the bismuth site has been noted to raise the T_c of the $Bi_2Sr_2CaCu_2O_8$ phase[10]. We have observed an increase in T_c from ~85 K for $Bi_2Sr_2CaCu_2O_8$ to ~97 K for $Bi_{2-x}Pb_xSr_2CaCu_2O_{8-x/2}$ (x = 0.5). The $Bi_2Sr_2CuO_6$ phase is made to superconduct at ~12 K by copper-rich off-stoichiometric melting (Table I). The substitution of lead into this phase, $Bi_{2-x}Pb_xSr_2CuO_{6-x/2}$ (x = 0.5), doubles the T_c to ~25 K. Electron microscopy has shown that the ~5x superstructure associated with the a-axis of the Bi/Sr materials is absent in the Pb substituted 221 material. We are now in the process of determining the detailed structure of these lead-doped compounds.

Structures Of The Tl/Ba Compounds

The structures[5,7] of $Tl_2Ba_2Ca_{n-1}Cu_nO_{2n+4}$ with n = 1 and 2 are very similar to those of the analogous Bi/Sr compounds discussed above, but possess higher crystallographic symmetry and fairly well-ordered thallium-oxygen double layers. Crystal structures for the Tl/Ba phases with n = 1, 2, and 3 are shown in Figure 2. The structures differ from one another by the number of consecutive Cu-O sheets. In $Tl_2Ba_2Ca_2Cu_3O_{10}{}^4$, triple sheets of corner-sharing square-planar CuO_4 groups are oriented parallel to the (001) plane. Additional oxygen atoms are located above and below the triple Cu-O sheets. As in the n = 2 Bi/Sr and Tl/Ba phases, there are no oxygen atoms between the triple Cu-O sheets. A small amount of thallium (~12 %) substitutes for the calcium found between the Cu-O layers of the n = 2 and 3 phases. In addition, it is not yet clear if the thallium layers are partially vacant, or if there is some calcium substitution (~15 %).

Barium ions are found above and below the Cu-O single, double, or triple sheets in nine-coordination with oxygen. These slabs alternately stack with a double thallium-oxygen layer along the c axis. Thallium bonds to six oxygen atoms in a distorted octahedral arrangement, where the octahedra share edges within and between each sheet of the double layer. A subtle disorder is observed in the Tl-O sheets, where the atoms are shifted from their ideal positions to create a more favorable bonding environment[4,5,7].

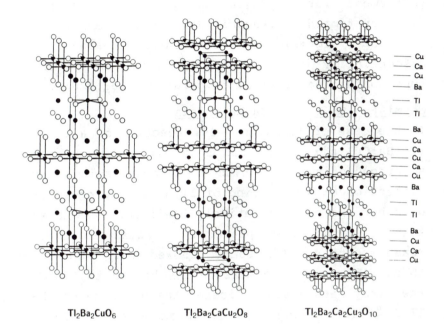

	Cu
	Ca
	Cu
	Ba
	Tl
	Tl
	Ba
	Cu
	Ca
	Cu
	Ca
	Cu
	Ba
	Tl
	Tl
	Ba
	Cu
	Ca
	Cu

$Tl_2Ba_2CuO_6$ $Tl_2Ba_2CaCu_2O_8$ $Tl_2Ba_2Ca_2Cu_3O_{10}$

FIGURE 2.   Structure of $Tl_2Ba_2CuO_6$, $Tl_2Ba_2CaCu_2O_8$, and $Tl_2Ba_2Ca_2Cu_3O_{10}$. Metal atoms are shaded and Cu-O bonds are shown.

The structures of $Tl_2Ba_2CaCu_2O_8$ and $Tl_2Ba_2Ca_2Cu_3O_{10}$ were also refined from neutron powder diffraction data[11]. Data for $Tl_2Ba_2CaCu_2O_8$ were taken at 298 K, and the data for $Tl_2Ba_2Ca_2Cu_3O_{10}$ were obtained at both 150 and 13 K. The results are essentially in agreement with the structural refinements based on single crystal x-ray diffraction data and confirm the positional disorder of the oxygen atoms in the Tl-O planes and the absence of oxygen between consecutive copper-oxygen sheets. The low-temperature refinements for $Tl_2Ba_2Ca_2Cu_3O_{10}$ indicate that the symmetry remains tetragonal down to 13 K, and there is no significant structural change or discontinuity in the cell parameters through the superconducting critical temperature near 125 K.

Chemistry Of The Tl/Ba Materials

In the case of Tl/Ba (n = 1), reaction of a 2:2:1 (Tl:Ba:Cu) mixture at 850°C in air resulted in a superconducting phase with a characteristic x-ray reflection at a d-spacing of 11.5 Å. The presence of other phases at relatively low levels was also seen in the XRD pattern. Flux exclusion measurements on this sample showed a T_c onset of 55 K. When the same mixture was heated to 875°C, the x-ray data again indicated the presence of the 11.5 Å line, along with small amounts of $BaCuO_2$. This material showed a T_c onset of 84 K and apparent zero resistivity at 79 K. On heating to 900°C, the 2:2:1 mixture partially melted, and the x-ray pattern showed large amounts of $BaCuO_2$ along with the superconducting phase, and small amounts of Tl_2O_3. Flux exclusion measurements on this product indicated a T_c onset of ~90 K, and electrical measurements showed apparent zero resistivity at 83 K (Figure 3).

For $Tl_2Ba_2CaCu_2O_8$ reactions containing excess CuO and CaO, powder x-ray diffraction revealed that most products were mixtures containing mainly two phases with characteristic layer spacings of ~14.7 Å and ~18 Å. Flux exclusion measurements showed superconductivity with onset temperatures

228

SYNTHESIS OF $A_2B_2Ca_{n-1}Cu_nO_{2n+4}$ SUPERCONDUCTORS
ranging from 100 to 127 K. The samples with an
onset of 127 K are poorly crystalline, and the
presence of a 2223 phase is revealed by the x-ray
reflection at a d-value close to 18 Å. The 2212
phase appears in most preparations, and the highest
transition temperatures are observed when the
samples are partially melted above 900°C. Flux
exclusion measurements on the 2212 crystals
revealed a sharp superconducting transition at T_c
= 110 K. Figure 3 shows the resistance for a
polycrystalline sample of $Tl_2Ba_2CaCu_2O_8$ that also
contained minor amounts of other phases. However,

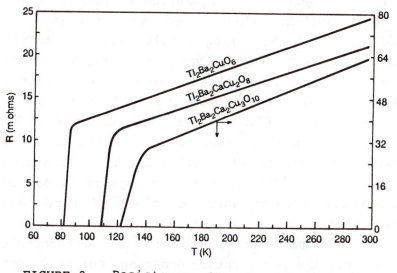

FIGURE 3. Resistance data for Tl-Ba-(Ca)-Cu-O
superconductors.

the 2212 compound can be made cleanly by starting
with the stoichiometric quantities of oxides as
given in Table I.

Flux exclusion measurements obtained on the

$Tl_2Ba_2Ca_2Cu_3O_{10}$ preparations showed that the superconducting transitions varied widely (105 to 127 K), depending on the synthesis conditions. Four-probe resistivity measurements on a typical sample showed a drop in the resistivity at ~140 K and zero resistivity at 122 K[4] (Figure 3). Flux exclusion measurements on the same sample showed an abrupt transition around 125 K, as observed by others[12]. The resistivity drop at 140 K is believed to be due to the n = 5 intergrowths observed in electron microscopy lattice imaging[4].

Comparisons Of The Bi/Sr and Tl/Ba Materials

For the $A_2B_2Ca_{n-1}Cu_nO_{2n+4}$ series of compounds, we have established that n may reach values of 2 (Bi/Sr) and 3 (Tl/Ba) as a bulk phase. Although bulk materials have not yet been prepared for Bi phases with n greater than two, or for Tl phases with n greater than three, we find evidence for intergrowths with n = 3 for Bi/Sr and n = 5 for Tl/Ba. It is thus tempting to conclude that the onset behaviour at about 117 K in the Bi/Sr system, and at about 140 K in the Tl/Ba system, is due to these intergrowths, where n values of three and five, respectively, are achieved.

The bismuth-strontium phases are orthorhombic and exhibit superstructure ordering, but we do not yet know the effect of such superstructure on the superconducting properties. In contrast, the thallium-barium compounds are tetragonal and do not show any obvious superstructure, although a subtle disorder, manifested in a locally distorted

SYNTHESIS OF $A_2B_2Ca_{n-1}Cu_nO_{2n+4}$ SUPERCONDUCTORS

Tl environment may actually exist.

The nature of the Bi-O versus Tl-O bonding within the double layers is responsible for the different morphologies of the two materials[5,7]. The mica-like character of the Bi/Sr compounds is due to weak intersheet Bi-O bonds. In these materials, the distance between adjacent bismuth sheets is ~3.25 Å. Cleavage between the Bi-O layers yields charge neutral segments. The interaction between the neighboring Tl-O sheets in the three Tl/Ba compounds is much stronger. Intersheet Tl-O bonds are ~2.0 Å in length and the separation between the two Tl layers is 2.2 Å which makes the crystals more difficult to cleave. The difference in the intersheet bonding of the Bi-O and Tl-O double layers is reflected in the c axis length of the n = 1 and 2 compounds. Although the average ionic radii of Bi/Sr (1.17 Å) and Tl/Ba (1.18 Å) are almost identical, the c axes of the bismuth phases are ~1.4 Å (n = 1) and ~1.6 Å (n = 2) longer than those of the corresponding thallium phases.

REFERENCES

1. H. Maeda, Y. Tanaka, M. Fukutomi, and T. Asano, Jap. J. Appl. Phys., 27, L209 (1988).
2. C. W. Chu, J. Bechtold, L. Gao, P. H. Hor, Z. J. Huang, R. L. Meng, Y. Y. Sun, Y.Q. Wang, and Y. Y. Xwe, Phys. Rev. Lett., 60, 941 (1988). 3.Z. Z. Sheng and A. M. Hermann, Nature, 332, 138 (1988).
3. Z. Z. Sheng and A. M. Hermann, Nature, 332, 138 (1988).

4. C. C. Torardi, M. A. Subramanian, J. C. Calabrese, J. Gopalakrishnan, K. J. Morrissey, T. R. Askew, R. B. Flippen, U. Chowdhry, and A. W. Sleight, Science, 240, 631 (1988).
5. C. C. Torardi, M. A. Subramanian, J. C. Calabrese, J. Gopalakrishnan, E. M. McCarron, K. J. Morrissey, T. R. Askew, R. B. Flippen, U. Chowdhry, and A. W. Sleight, Phys. Rev. B., submitted.
6. M. A. Subramanian, C. C. Torardi, J. C. Calabrese, J. Gopalakrishnan, K. J. Morrissey, T. R. Askew, R. B. Flippen, U. Chowdhry, and A. W. Sleight, Science, 239, 1015 (1988).
7. M. A. Subramanian, J. C. Calabrese, C. C. Torardi, J. Gopalakrishnan, T. R. Askew, R. B. Flippen, K. J. Morrissey, U. Chowdhry, and A. W. Sleight, Nature, 332, 420 (1988).
8. B. Grande, H. Müller-Buschbaum, and M. Schweizer, Z. Anorg. Allgem. Chem., 428, 120 (1977).
9. C. C. Torardi, E. M. McCarron, H. S. Horowitz, J. B. Michel, A. W. Sleight, and D. E. Cox, in Chemistry of High-Temperature Superconductors, edited by D. L. Nelson, M. S. Whittingham, and T. F. George (American Chemical Society, Washington, 1987), p. 152.
10. S. A. Sunshine, T. Siegrist, L. F. Schneemeyer, D. W. Murphy, R. J. Cava, B. Batlogg, R. B. van Dover, R. M. Fleming, S. H. Glarum, S. Nakahara, R. Farrow, J. J. Krajewski, S. M. Zahurak, J. V. Waszczak, J. H. Marshall, P. Marsh, L. W. Rupp, Jr., and W. F. Peck, Phys. Rev. Lett., submitted.
11. D. E. Cox, C. C. Torardi, M. A. Subramanian, J. Gopalakrishnan, and A. W. Sleight, Phys. Rev., submitted.
12. S. S. P. Parkin, V. Y. Lee, E. M. Engler, A. I. Nazzal, R. Beyers, T. C. Huang, G. Gorman, and R. Savoy, Phys. Rev. Lett., submitted.

WHAT'S HOT IN SUPERCONDUCTIVITY AT NRL

S. A. WOLF
Materials Physics, Naval Research Laboratory, Washington,
DC 20375-5000, USA

Abstract.   I will review the latest results on the preparation and
properties of the high superconducting transition temperature
materials including some recent results on the Bi-Sr-Ca-Cu-O
materials. Bulk, single crystal, and thin films will be discussed.

SYNTHESIS AND CHARACTERIZATION OF BiCaSrCuO AND

BiSnCaSrCuO SUPERCONDUCTING CERAMICS

JOHN H. MILLER JR.[b], WILLIAM E. HATFIELD[a], BRIAN R.
ROHRS[a] MARTIN L. KIRK[a], JOANNA L. PERKINSON[a], KATHLEEN
L.TROJAN[a], JOHN D.HUNN[b], ZU HUA ZHANG[b], WILLIAM J.
RILEY[b]
The Departments of Chemistry[a] and Physics[b], The
University of North Carolina at Chapel Hill, Chapel
Hill, North Carolina 27599, U.S.A.

Abstract   Magnetic susceptibility measurements,
resistance measurements, Rutherford backscattering,
scanning electron microscopy, and energy dispersive
X-ray analyses were utilized in order to determine the
effects of composition and firing conditions on the
properties of the Bi-Ca-Sr-Cu-O superconducting
ceramics.

INTRODUCTION

New high - temperature superconducting materials with T_c's
in the range of 120K have recently been realized in both
the Tl-Ca-Ba-Cu-O[1] and Bi-Ca-Sr-Cu-O[2] systems. These
systems consist of at least two superconducting phases
characterized by their superconducting transition
temperatures of ~80K and ~120K. The transition
temperature, · critical current density, anisotropic
properties, and volume fraction of the superconducting

material are inherently related to the microstructure of the ceramics. In this report, we have investigated the effects of starting composition, Sn doping, and annealing and sintering conditions on the magnetic and electrical properties, microstructure, surface stoichiometry, and multiphasic nature of the Bi-Ca-Sr-Cu-O system.

EXPERIMENTAL

Synthesis

Samples of the Bi-Ca-Sr-Cu-O ceramics were synthesized via intimate mixing of Bi_2O_3, CuO, $CaCO_3$, and $SrCO_3$ by thoroughly grinding the starting materials. This was followed by pressing into a pellet and sintering at $880^\circ C$ (sample 1), doping with SnO and SnO_2, pressing into a pellet, sintering at $880^\circ C$, and annealing at $950^\circ C$ for 15 minutes (sample 2), pressing into a pellet, and sintering at $820^\circ C$ (samples 3 and 4). All materials were sintered and annealed in air in a muffle furnace. The ratios of the starting materials are given in Table 1.

TABLE I Molar ratios of starting materials

Sample	Bi-Sn-Ca-Sr-Cu molar ratios
1	2-0-1-2-2
2	2-1-1-2-2
3	1-0-1-1-2
4	2-0-1-2-2

Bi(Sn)CaSrCuO SUPERCONDUCTING CERAMICS

Magnetic Measurements

Quantitative magnetic measurements were made using a vibrating sample magnetometer equipped with a continuous flow cryostat. The samples were cooled in zero field (ZFC) to 4.2K, and the diamagnetic shielding data were collected past the transition temperature in a field of 50 Oe. The sample was then field-cooled (FC) through the transition temperature in order to monitor the Meissner effect. The density of the samples was measured and volume susceptibilities were determined. All data were corrected for demagnetization effects.

Resistance Measurements

Resistance measurements were made from 20 to 300K utilizing an in-line four-probe technique. Compensation for the Seebeck effect was made.

Surface Analysis

The surface, microstructure, and stoichiometry of the superconducting ceramics were determined by Rutherford backscattering (RBS) and scanning electron microscopy - energy dispersive X-ray analysis (SEM-EDX). A standardless semi-quantitative analysis program was used to determine the elemental analysis of the bulk material as well as that of the individual grains.

RESULTS AND DISCUSSION

Transition Temperature and Superconducting Volume Fractions

The magnetic data for the four samples studied show

relatively broad transitions, indicative of the multiphase nature of the samples. Sample 1 displayed the greatest volume fraction of superconducting material, with 8.4% bulk diamagnetism in the Meissner signal and 13.1% for the maximum diamagnetic shielding. The T_C for sample 1 was monitored by both resistance and magnetic measurements. The onset temperature of the decrease in resistance was 87K, with zero resistance being realized at 37K. The magnetic data determined the T_C to be 77.6K for the ZFC run and 79.5K for the FC data set. The anomalously low temperature necessary for zero resistance is most probably due to contact problems or the percolative nature of the superconductivity. The results of the critical temperatures and diamagnetic fractions are summarized in Table II.

Surface Analysis and Microstructure

RBS of samples 1 and 2 yielded valuable information concerning the constitution of the sample surface. The surface stoichiometries (Bi-Sn-Sr-Ca-Cu) were determined to be (1:0.0:0.7:1.0:1.5) and (1:0.53:1.27:0.93:1.33) respectively. This data indicates that the surface is rich in Ca and Cu, with the latter probably being present in the form of oxides.

An EDX elemental analysis of sample 1 yielded a stoichiometry of $Bi_{1.43}Sr_{1.62}Ca_{0.61}Cu_{2.00}$ and the SEM showed the material to be very porous, with little or no melting. Analysis of single grains confirmed the multiphasic nature of the composite, however some grain of different morphology displayed very similar stoichiometries.

238

TABLE II   Transition temperatures and percent bulk
            diamagnetism

Sample	T_c(ZFC)	T_c(FC)	T_c(onset)	T_c(zero)	%(ZFC)	%(FC)
1	77.6	79.5	87	37	13.1	9.4
2	72.3	76.9	88	61	8.2	1.6
3	69.5	72.7	--	--	5.6	4.1
4	73.5	73.5	--	--	1.0	0.7

The overall stoichiometry of sample 2 was
$Bi_{0.61}Sn_{0.55}Sr_{1.28}Ca_{0.90}Cu_{2.00}$. Analysis of single grains
showed large areas of Sr depletion. As in the case of
sample 1, no melting was observed, and the individual grains
were of widely differing morphology and composition.

Sample 3 was very homogeneous, and individual grains had
stoichiometries similar to that of the bulk
($Bi_{1.50}Sr_{0.91}Ca_{1.07}Cu_{2.00}$). However, we did find areas
heavily depleted in Bi, Sr, and Ca. In contrast to the
previous ceramics, SEM showed that some melting had occurred,
resulting in partial fusion of the grains in an annealed
pellet. Analysis of some unannealed sintered powder yielded
some lamellar grains of composition $Bi_{1.7}Sr_{1.4}Ca_{0.9}Cu_{2.0}$.

EDX of the final sample indicated severe Ca depletion.
The sample was polycrystalline and very homogeneous (1μ
dia.). As in the case of the first two ceramics, the
pressed pellet was very porous.

CONCLUSIONS

The samples studied were generally multiphase in nature and very porous. The transition temperatures were found to be dependent on the molar ratios of the starting materials and firing conditions, but generally, the T_c's correspond to the 80K phase. More intimate mixing of the precursors must be obtained in order to increase the homogeneity of the fired material and increase the yield of the superconducting fraction. It may be that the best reaction conditions for the highest yields occur in a very narrow temperature range, requiring precise control of the firing temperature.

ACKNOWLEDGEMENTS

This work was supported in part by the Office of Naval Re search. J.H.M. acknowledges the receipt of an A.P. Sloan Research Fellowship, an R.J. Reynolds Faculty Development Award, and a grant from the U.N.C. Research Council.

REFERENCES

1. a) Z. Z. Sheng and A. M. Hermann, _Nature_, _332_, 55 (1988).

   b) Z. Z. Sheng and A. M. Hermann, _Nature_, _332_, 138 (1988).

2. a) H. Takahi, H. Eisaki, S. Uchida, A. Maeda, S. Tajima, K. Uchinodura and S. Tanaka, _Nature_, _323_, 236 (1988).

   b) R. M. Hazen, C. T. Prewitt, R. J. Angel, N. L. Ross, L. W. Finger, C. G. Hadidiacos, D. R.Veblen, P. J. Heaney, P. H. Hor, R. L. Meng, Y. Y. Sun, Y. Q. Wang, Y. Y. Xue, Z. J. Huang, L. Gao, J. Bechtold, and C. W. Chu, _Phys. Rev. Lett._, _60_, 1174 (1988).

SUPERCONDUCTING OXIDE PROCESSING IN THE Bi-Ca-Sr-Cu-O SYSTEM: CHARACTERIZATION AND COMPARISON BY AEM

C. M. SUNG, D. HONG, M. P. HARMER, and D. M. SMYTH
Department of Materials Science and Enginering, Materials Research Center, Lehigh University, Bethlehem, PA 18015, USA

Abstract. Microstructures of superconducting phase with transitions near 85 K and 110 K have been studied by analytical electron microscopy (AEM) in samples of the system Bi-Ca-Sr-Cu-O prepared by sintering and liquid mixing methods. The heat treatment was restricted to a temperature range near 860°C to obtain good superconducting behavior. The microchemistry and the crystal structure of the superconducting phase in the Bi-Ca-Sr-Cu-O system were investigated by energy dispersive x-ray spectroscopy and convergent-beam electron diffraction, and were related to the processing routes.

PROCESSING AND CHARACTERIZATION IN THE Bi-Sr-Ca-Cu-O SYSTEM

DARLA SCHRODT, BARRY BENDER, STEVEN LAWRENCE,
ROY RAYNE, MIKE OSOFSKY, VALERIE LETOURNEAU,
WILLIAM LECHTER, ANAND SINGH, HENRY HOFF,
LLOYD RICHARDS, EARL SKELTON, SYED QUADRI,
CHANDRA PANDE, ROBERT SOULEN, STUART WOLF, and
DON GUBSER
Naval Research Laboratory, Washington, DC 20375 ,USA

Abstract. The new Bi-Sr-Ca-Cu-O high-T_c superconductor has been processed using a variety of techniques with varying degrees of success. Characterizations of the resulting samples include microstructural analysis using optical microscopy and secondary electron microscopy, compositional analysis using energy dispersive spectroscopy, structural analysis using transmission electron microscopy and x-ray diffractometry, and electromagnetic analysis using four-probe AC (25.5 Hz) resistance measurements and magnetization in a 200 G field. Results and correlations are presented.

MICROSTRUCTURAL CONSIDERATIONS IN POLYCRYSTALLINE YBa$_2$Cu$_3$O$_7$

R. W. MCCALLUM, J. D. VERHOEVEN, AND
A. J. BEVOLO

Ames Laboratory and Department of Materials Science and Engineering,
Iowa State University, Ames, Iowa 50011

Abstract The microstructure of sintered bulk samples of YBa$_2$Cu$_3$O$_7$ has readily observable effects upon both the transport properties and the DC magnetization. Following a field change or a rapid temperature change, a time-dependent relaxation is observed in the magnetization, which is directly related to the weak links joining the superconducting grains. The functional form of the relaxation is related to the critical current of these links. Scanning Auger Microscopy of fracture surfaces indicates that grain boundaries may be severely strained , and may be subject to environmental attack. The weak-link behavior appears to be related to strain and microcracking of the grain boundaries.

INTRODUCTION

The major obstacles to the application of high-temperature superconductors are the extremely low critical currents obtained for polycrystalline sintered materials and the extreme brittleness of these ceramic materials. Scanning Auger Microscopy measurements and studies of grain growth as a function of temperature indicated that these two phenomena are related and that the low critical currents result from microcracks which form during processing of the materials. The microcracked boundaries serve as weak links between grains. For large – grained material with random grain orientation, the large anisotropy in the thermal expansion of

$YBa_2Cu_3O_7$ results in strains at the grain boundaries which are sufficient to crack the boundaries. In order to limit cracking, very small grain size is required. While low-temperature sintering allows the retention of small grain size, sintering and densification is poor. In order to sinter at temperatures above 950°C without exaggerated grain growth, all impurity phases which melt below this temperature must be eliminated.

SAMPLE PREPARATION

Samples of $YBa_2Cu_3O_7$ were prepared by mixing and grinding dried powders of $BaCO_3$, Y_2O_3, and CuO. Samples were ground in a motorized mortar and pestle for a period of 1 hour per 10 grams of material. The material was pelletized and heated in air for 24 hours at 890°C, after which it was reground and pelletized before a second 24 hours at 890°C in air and a final grinding and pelletizing. For the Auger experiment the material was then heated to 1000°C for 3 days in flowing O_2. For the grain growth experiments the temperature of this final sinter was varied as discussed below. Samples were cooled at 3.5°C/min, held at 500°C for 1 h, and then furnace cooled. Field-cooled measurements in a DC magnetometer gave a T_c onset at 92K with a sharp rise to a maximum Meissner expulsion of 30 to 45% at 85 K for all samples. The Meissner signal varied systematically with the sample grain size as expected.

MAGNETIZATION RELAXATION

The weak-link behavior of the superconducting state is easily observed in these materials by studying the relaxation of the DC magnetization following a field change. When the field through a superconducting loop containing a weak link is changed, a current opposing this change is produced. Due to flux flow in the weak link, this current decays as a function of time. If the induced current is much less than the critical current of the link, this decay is logarithmic in time. This behavior has been demonstrated by Mota, et al[1] for the oxide superconductors. If the induced current exceeds J_c of the link, then the decay is exponential. For values close to J_c the relaxation has a non-analytical form. We have investigated the relaxation of the DC magnetization under two extreme conditions. The first is low temperatures for large (1 T) field changes and the second is for $T > .8\ T_c$ and field changes ranging from 2 to 60 mT. In both cases the relaxation is neither logarithmic nor exponential. While these results suggest that in both cases the induced currents are of the order of the critical current of the weak links, it is possible that the exponential part of the relaxation occurs so rapidly that we do not observe it due to the slow speed of our magnetometer measurements. It is however unmistakable that we observe weak link type behavior.

SCANNING AUGER MICROSCOPY

Two samples of $YBa_2Cu_3O_7$ measuring approximately 2 mm x 3 mm x 8mm were cut from the center of a single sintered pellet. It should be noted that this pellet had been exposed to ambient atmosphere for a period of months. One of the samples was polished on one side and micrographs were made of the polished surface. Following in–situ fracturing in the Auger chamber, these micrographs were used to unambiguously determine the type of fracture which occurred. Details of this experiment have been presented elsewhere[2]. Of a total of 63 grains along the 2.8 mm-wide polished edge, 24 fractured transgranularly or partially transgranularly, and several cracks were along preexisting cracks rather than a result of cleavage. The sample fractured predominantly along planar boundaries, pores and cracked grains. C Auger maps showed that these surfaces are all C-contaminated, at least after polishing. The thickness of the carbon layer was found to be one monolayer.

Considerable effort was spent evaluating the C level on the two types of pores. A clear correlation was found: internal pores were consistently free of C contamination, whereas interface pores were consistently covered with C.

Based on these results, it was concluded that the C contamination might have occurred during sample handling and polishing, and not during the actual processing. If the grain boundaries were sufficiently open,perhaps because of cracking, it is possible that they could have been contaminated by the liquids used in polishing and cleaning,or perhaps even by CO_2 in the air. This is consistent with the absence of C on the internal pores and transgranular surfaces, and the higher C contamination on grain boundaries near the polished surface. To check this possibility, the second sample was heated to 500°C in flowing O_2 for 16 hr and then furnace cooled. The sample was handled so as to limit the exposure to air to 20 seconds during the transfer to the Auger spectrometer. The C/O ratio immediately after fracture was decreased by a factor of 20 from that found on the fracture surface of the first sample. This indicates that the C buildup on the grain boundary surfaces of the first sample was not produced in the original processing, but resulted from subsequent environmental contamination on handling. It appears that a majority of the grain boundaries in the large-grained samples prepared here are sufficiently "open" that they become contaminated during standard metallographic preparation.

GRAIN GROWTH STUDIES

If the grain boundaries of large-grained samples are cracked, this almost certainly arises from stresses associated with the large anisotropy in the thermal expansion of the material. In order to minimize the total stress on any given boundary, it is necessary to obtain very small-grained material so that the stress is distributed over a large number of grain boundaries.

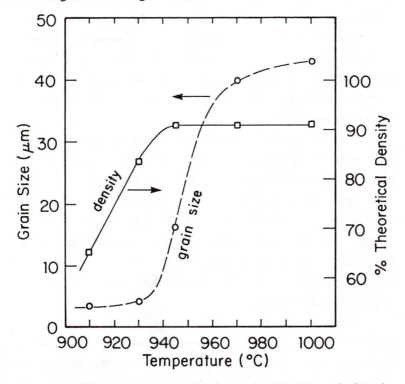

FIGURE 1. Grain size and density as a function of sintering temperature.

In Figure 1 we show the density and grain size for a series of samples which were prepared from the same batch of calcined powder by sintering in O_2 at different temperatures. There is a step in the grain size at approximately 940°C. For samples processed at higher temperatures the grain size is essentially constant, and there is no discernible difference between samples processed for 24 hrs at temperature and those held 72 hrs. Differential thermal analysis of the calcined starting material shows the presence of a melting endotherm at this temperature. Thus, as has been postulated in a previous study, exaggerated grain growth takes place as a result of liquid-phase sintering. This dependence on a small quantity of liquid phase means that the microstructure one obtains following sintering at a given temperature is highly dependent on the phases present at that temperature. This in turn is highly dependent upon the method used to prepare the calcined powders. For liquid-phase sintering to be effective, only a percent or so of the sample must be liquid. This is well below the limits of detection by most conventional means in these complex materials.

The density of the samples shows an increase at a slightly lower temperature than the grain size, so that there is a very narrow band in which the density is reasonable, while the grains are small. This region should give optimum critical currents, and indeed Inuzuka et al.[4] observe a peak in the critical current in such a narrow region. It should be noted that their grain growth takes place at 900 C which suggests that their samples contain the $BaCuO_2$ - CuO eutectic rather than the $BaCuO_2$ which appears to be present in our samples.

These results have two important consequences. First, in order to obtain high density, fine-grained materials, it is extremely important to avoid low-melting impurity phases in the samples. Even as little as one percent of such a phase will lead to exaggerated grain growth above its melting temperature. Second, due to the extreme dependence of the microstructure on impurity phases, substitution experiments are extremely difficult, since even if the substituted element goes primarily into the 123 lattice, if it affects the melting temperature of impurity phases or the reaction temperatures for calcining, the effects on the microstructure may outweigh any substitutional effects.

CONCLUSIONS

The critical currents of high-T_c superconductor YBa$_2$Cu$_3$O$_7$ are limited by weak-link behavior, which arises from strained and microcracked grain boundaries. In order to eliminate these defects, fine-grained material must be produced. Since exaggerated grain growth takes place in the presence of liquid phases, all impurity phases which are liquid at the processing temperature must be avoided.

ACKNOWLEDGEMENTS

The authors would like to thank S. B. Luethje, E. D. Gibson, M. A. Noack, and K. No for laboratory assistance with these measurements. This work was performed at the Ames Laboratory, Iowa State University and was supported by the Director of Energy Research, Office of Basic Energy Science, U. S. Department of Energy under contract No. W-7405-ENG-82.

REFERENCES

1. A.C. Mota, A. Pollini, P. Visani, K.A. Muller, and J.G. Bednorz, Phys. Rev. B, 36(7), 4011 (1987).
2. R.W. McCallum, J.D. Verhoeven, M.A. Noack, E.D. Gibson, F. C. Laabs, D. K. Finnemore, and A.R. Moodenbaugh, Adv. Ceram. Mater. 2, 388 (1987).
3. J.D. Verhoeven, A.J. Bevolo, R.W. McCallum, E.D. Gibson and M.A. Noack, Appl. Phys. Lett 52(9), 745 (1988).
4. T. Inusuka, T. Ando, A. Hayashi, K. Sawano, and H. Kubo Amer. Ceram. Soc. May 1988, 73-SII-88.

APPLICATION OF PRINCIPLES OF PROCESS METALLURGY TO THE PRODUCTION OF COPPER-SHEATHED Y-Ba-Cu-O WIRES

B. YARAR, E. L. BROWN, J. U. TREFNY, N. MITRA, AND G. PINE
Colorado School of Mines, Golden, Colorado

Abstract Principles of process metallurgy were used in
the preparation of Y-Ba-Cu-O powders, their purification
by magnetic methods, characterization by various physical
and metallurgical techniques including the use of a
"superconductivitymeter" and cold rolling of
powder-packed tubes for wire manufacture. It was found
that while a "mini" grinding process takes place during
wire drawing, the limiting size of a wire is controlled
by strain relationships between the core and the copper
sheathing. Temperature effects on core material
properties are also reported.

INTRODUCTION

We have been involved in high-Tc superconductor materials
science and technology since we have reproduced the
La-Ba-Cu-O materials in early 1987. We have since
carried out research with La-Ba, Eu-Y, Y-Ba containing
materials, and more recently we are investigating
rare-earth-free materials, such as Tl-compounds [1-3] and
Bi-containing products.

Our engineering efforts have produced discs, rings,
rods, and various monolithic shapes, as well as thin films
and polymer composites. We report in this paper part of
our effort to elucidate the feasibility of the
manufacture of wires by the powder-in-the-tube-type
method, which has been described in the literature.

253

POWDER-IN-THE-TUBE METHOD AND SUPERCONDUCTOR WIRE
PRODUCTION

This technique also known as "deformation processing",
which includes rolling, swagging and drawing, consists of
filling a suitable tube with superconducting powder,
sealing both of its ends, and rolling or drawing it to a
wire.

The technology for the manufacture of wires and
filaments from conventional superconducting alloys has
been outlined by Roberge [4], where numerous examples of
filaments with diameters of 1 to 3 microns are given.
Methods of production include the copper sheathing
approach as well. The principle of this approach is, in
fact known as the Kunzler Method [5], where a niobium
tube with internal and external diameters of 3 and 6 mm,
respectively, is packed with Sn, and Nb in the appropriate
proportion to give Nb_3Sn and the sealed tube is drawn as
shown in Figure 1. The product has 0.38 mm OD and after
annealing at 1000°C can carry up to 1.5×10^5 A/cm^2,
which is higher than that of the bulk material. This
approach has also been cited for the High-Tc Y-Ba-Cu-O
compounds as potentially suitable. McCallum et. al. [6]
for example, have given some "speculative ideas" on
various techniques which include the powder in the tube
method. Jin et. al. [7], on the other hand, report the
use of Ag alone, or Cu with a Ni/Au diffusion barrier, and
the production of wires from 2.5 mm to 0.25 mm diameter.
These authors note that the T_C values of the core
material remain unaltered after the drawing process, and
indicate that the new wire can carry about 175 A/cm^2 at
77K. One trip report [8] cites that the Sumimoto

254

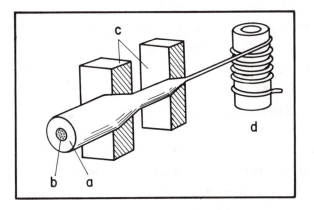

Figure 1 – The Kunzler method of superconductor wire
production (schematic). (a) copper sheath, (b) core
powder, (c) draw plates, (d) solenoid. After ref. 14, by
permission.

Electric Company in Japan has Y–Ba–Cu–O–based wire where
the maximum density reached is 90% of theoretical and
$J_c = 1240 \ A/cm^2$.

Other techniques applied to the high–Tc 123
materials which have been reported include short
extrusion products, tape casting, melt spinning, and
explosively formed sandwiches of the material between
copper sheaths [9–11,15]. Extrusion and casting products
need to be annealed for oxygen stoichiometry and removal
of binder material by combustion, which result in
products that require the improvement of their mechanical
properties. Although the explosively fabricated products
can be suitable for certain applications, such as
rf–shielding, the continuity of the sandwiched part and

255

its critical current have not yet been unequivocally established.

The study reported in this paper was undertaken to establish the feasibility or limitations of the deformation processing method using 123 ··Y-Ba-Cu-O powders and copper sheathing.

EXPERIMENTAL PROCEDURES

The overall scheme of our operation, given in Figure 2, consisted of the following components: powder preparation, multi-stage magnetic separation, and tube preparation, packing and drawing with appropriate characterizations at all stages.

Solid State Reaction-Preparation of the 123 Powder

This well-established method allows the production of superconducting powders which exhibit reproducible Tc and X-ray diffraction characteristics. Although our superconductor preparation methods include hydrolytic co-precipitation and organic precursor approaches, the data presented in this paper were generated with products of solid state reactions.

In this procedure problems may arise when the material is produced in large quantities. Here, the limited diffusion of oxygen to the powder bed produces only partial conversion to the desirable orthorhombic structure, the rest of the powder remaining either unreacted or oxygen-deficient. The same is observed when high-temperature treatment steps are kept short. In some cases reactor materials or grinding media, especially if ball milling is used for size reduction, contaminate the

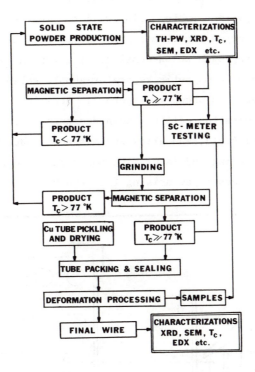

Figure 2 - Flowsheet of processes and characterizations.

product. The contaminating, non-superconductive fractions can be separated by magnetic methods widely used in mineral processing technology. Physical powder separation techniques exploit physico-chemical differences between materials, such as light or radiation absorption (optical sorting), density (gravity separation), surface chemistry (flotation, flocculation), and magnetic property differences [12]. The last property is readily noticeable as the Meissner effect in superconductors. We have been investigating the separation of powder components which exhibit the Meissner

257

effect and those that do not. While the full details of
our investigations on this line will be reported in a
separate paper, we report here the use of a magnetic
separation approach, which consists of a conventional
Franz Isodynamic magnetic separator, supplemented with
suitable parts, as shown in Figure 3. Here it is seen
that the separation cell has been designed to concentrate
the magnetic field, and allowance has been made, by
placing operating and insulating boxes inside one another,
to minimize turbulence arising from the boiling of liquid

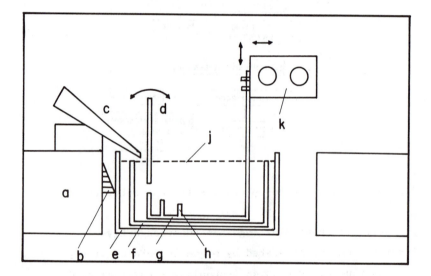

Figure 3 – Magnetic separation apparatus for the
separation of particles showing Meissner effect
(schematic). (a) magnet pole, (b) magnetic field
orientation, (c) vibratory feeder, (d) deflector plate,
(e) outer box, (f) inner box, (g) superconductor
collection chamber, (h) chamber partitions, (j) liquid
nitrogen level, (k) vernier arrangement for collection
chamber positioning. Arrows show direction of motion.
Supports and electrical controls are not shown.

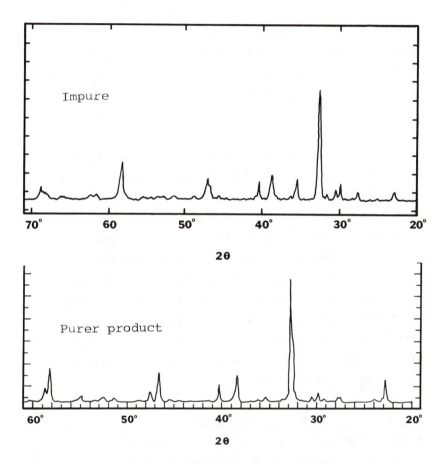

Figure 4 – Comparison of the XRD patterns of impure and magnetically separated superconductor fractions.

nitrogen during the separation operation. Figure 4 shows XRD data of impure and pure products obtained from this procedure. These materials also show differences in the magnitude of their Meissner Effect, as determined by the superconductivitymeter described below.

259

The Superconductivitymeter

This device was constructed for a quick quantification of the Meissner effect [13]. While our current laboratory capabilities include magnetic susceptibility determinations, thermo-electric power measurements, and other characterization techniques, this apparatus served as a preliminary testing device. The highlights of this instrument are shown in Figure 5. The samples are placed in a basket holder, which is placed in liquid nitrogen, or a medium where temperature determinations can be directly made. The force (F) exerted by means of an external magnetic field, such as one obtained by an electromagnet, is compensated for by an electronic feedback mechanism shown in this figure as "signal nullifier". The sample, therefore, does not move, but a signal, as shown in the insert of Figure 5, is obtained as a measure of the force applied. The signal-monitoring device is usually an x-y recorder. If the material contains a number of phases possessing different levels of purity or Tc values, the signal comes out stepwise indicating their presence.

Tube Packing and Cold Drawing Procedure

A copper tube was pickled in 5% v/v nitric acid for 5 minutes, washed with water and alcohol, and dried with pressurized air. One end of the tube was sealed by crimping, then packed with superconductive powder and sealed at the open end, followed by cold rolling into wire, using a two-high mill equipped with circular cross-section grooved rolls. Elongation per pass and wire OD were recorded after each pass. It was observed that a number of redundant passes through the same groove

260

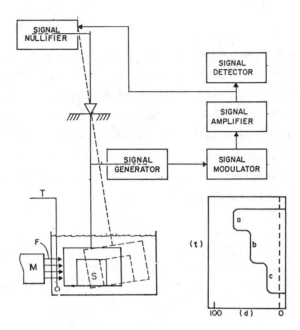

Figure 5 – Superconductivitymeter and components
(schematic): (M) magnet, (F) force, (T) thermometer, (S)
sample. Inset: superconductivitymeter signal, (t) time
axis, (d) deflection units on signal detector, (a) purest
sample, sample purity: a > b > c.

reduced flashing of the outer material, and gave a
smoother product. Samples were cut after each pass for
subsequent characterization procedures. The OD of the
initial tube was reduced from 6.35 mm to 0.88 mm in 15
passes. Specimen characterizations included SEM/EDX
analyses, monitoring of continuity by voltage-current
curves, core density determinations of specimens, and
critical current determinations of the finer cores.

WIRE CHARACTERIZATION RESULTS

Strain and Morphology

Specimens cut from various rolling stages were prepared
by mounting in a room temperature setting resin, followed
by standard grinding and polishing to a 0.05 micron
alumina finish. The specimens were then observed in a
JEOL 840 SEM equipped with a Tracor-Northern 5500/5600
x-ray analysis and image processing system.

Strain calculations were made from measurements on
electron micrographs as given below.

$$\% \ \varepsilon_{overall} = \frac{d_n^o - d_{n+1}^o}{d_n^o}$$

where:

d_n^o = overall diameter after nth roll pass

d_{n+1}^o = overall diameter after (n+1)th roll pass

$$\% \ \varepsilon_{core} = \frac{d_n - d_{n+1}}{d_n} \cdot 100$$

where:

d_n = diameter of core after nth roll pass

d_{n+1}^o = diameter of core after (n+1)th roll pass

$$\% \; \varepsilon_{sheath} = \frac{\Delta d_n - \Delta d_{n+1}}{\Delta d_n} \cdot 100$$

$\Delta d_n = d_n^o - d_n$ = difference between overall diameter and core diameter after nth roll pass

$\Delta d_{n+1} = d_{n+1}^o - d_{n+1}$ = difference between overall diameter and core diameter after (n+1)th roll pass

Figure 6 shows the evolution of strain per pass and cumulative strain the specimens undergo in 15 passes. It is seen in this figure that pass 13, which corresponds to a diameter reduction of about 80% of the original, the strain experienced by the core, undergoes a catastrophic drop. This indicates the following: "During the initial stages of rolling the overall wire deformation is distributed uniformly between the copper sheath and core. The sheath undergoes plastic deformation while the core densifies. At later stages, the core reaches a limiting degree of densification value, where essentially all of the deformation occurs in the sheath". At this stage the core cross-section ceases to be round and also undergoes cracking. Figure 7 shows cross-sections of the core at pass 10 (42.7% of original size) and pass 14 (25.4% of original size).

Particle Size Evolution During Core Densification
It has been intuitively suggested [7] that during the cold-rolling process core densification occurs by particles sliding against one another.

263

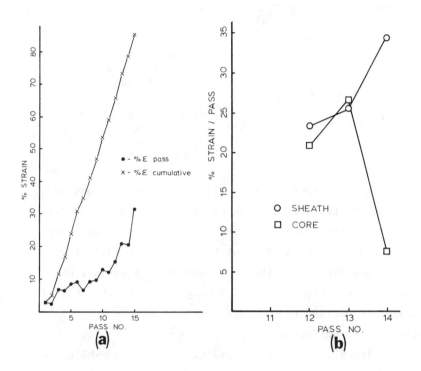

Figure 6 – Evolution of strain during the drawing process
(a) cumulative strains and (b) strain per rolling pass.

Our SEM observations indicate the following
phenomena to be occurring simultaneously during core
densification: (1) voidage is diminished by particle
movement, (2) cold compaction occurs, (3) particles
undergo a mini-comminution process, resulting with
multi-model particle size rearrangement, and (4) sheath,
core and core/sheath interface cracking occur in the
later stages of rolling, due to strain incompatibilities.

264

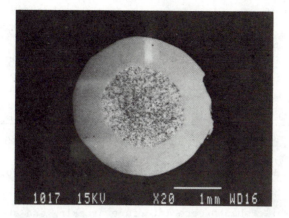

-10-

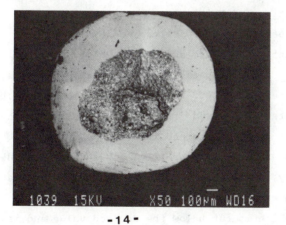

-14-

Figure 7 - Wire cross-sections at pass 10 and pass 14.

Processes 3 and 4, in particular, can be associated with the fact that copper is able to undergo plastic deformation, while the core material is not. The evolution of particle size distribution is shown in

265

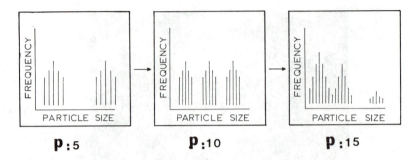

Figure 8 - Particle size evolution by progressive rolling.

Figure 8, while sheath/core interface cracking and sheath cracking are shown in Figure 9.

Core Densities and Critical Currents

Densities of 123 products, determined by the volume displacement method, were as high as 98% of the theoretical after preparation in the form of pellets and heat treatment at 950°C. Determinations made on sections of core material removed from rolled wires showed that the highest density obtained is 5.1 g/cc, which corresponds to 80% of theoretical, and indicates about 20% of porosity in the core. Critical current determination on the other hand, indicates a value of less than 10 A/cm^2, which is far below the desired value, and is believed to be a direct consequence of the low packing density, which accentuates the anisotropic effects known to be responsible for low critical currents observed in bulk materials. Intergranular resistances also contribute to the low critical currents observed.

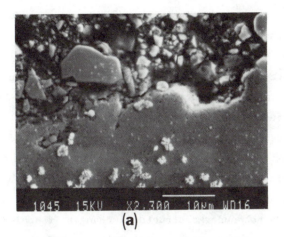

(a)

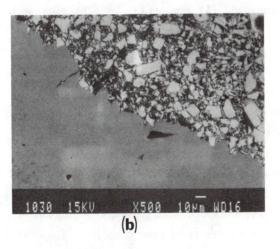

(b)

Figure 9 – Cracking of core and sheath during the later stages of rolling; (a) sheath/core interface cracking and (b) sheath cracking.

High-Temperature Behavior

Since it is known that 123 compounds cease to be superconductive at elevated temperatures, and require the

reintroduction of oxygen lost, experiments were conducted
to explore the potential for hot rolling. For this
purpose, superconductive pellets and cold-rolled wire
sections were subjected to various heating and cooling
cycles; their microstructures were studied by SEM and their
Meissner effect response was monitored by the superconductivitymeter. Oxygen contents of these pellets and
their XRD patterns were also monitored as a function of
heating regimes. Our findings in this series of
experiments are outlined below: (1) samples heated up to
$950 \pm 10°C$ do not undergo structural changes if cooled
slowly under an oxygen stream, (2) Meissner effect is
lost if samples are heated to 550°C in the absence of
flowing oxygen, and (3) compaction by sintering occurs at
temperatures higher than 980°C, with a complete loss of
Meissner effect, and an orthorhombic perovskite structure.

Figure 10 shows the variation of oxygen content of
the structure as a function of heating temperature. Figure 11
compares the XRD patterns of samples heated up to 550°C
and one heated to 1100°C.

Figure 12(a) shows the formation of relatively large
prismatic crystals when a sample is heated to 1100°C and is
allowed to slowly cool in the furnace under flowing
oxygen, while Figure 12(b) shows the morphology of a
sample quenched in air from 1100°C to room temperature.
This sample did not show a Meissner effect, and its
structure is clearly not conducive to the conduction of
current. We conclude, therefore, that post-rolling
annealing or hot rolling are not alternative solutions to
the sheath/core strain incompatibilities or low
compaction densities so far observed in the production of
wire by this method.

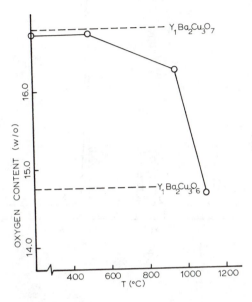

Figure 10 - Oxygen contents of core materials heated to various temperatures in air for one hour.

CONCLUSIONS

This study allows the following conclusions to be reached:

1)   The fabrication of wire by deformation processing of powder-packed copper tubes is limited by the low plasticity and fracture-toughness of the Y-Ba-Cu-O superconductor powder in the fully compacted state.

2)   In the early stages of rolling the strain in the core is associated with densification and particle motion.   Strain in the copper sheath is associated with plastic deformation.   Therefore, in the rolling process,

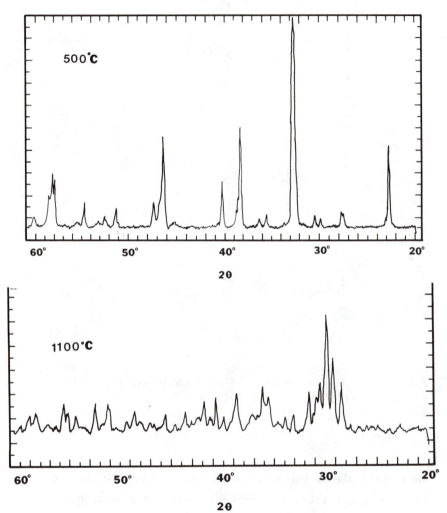

Figure 11 – XRD patterns of core material samples heated
at 500°C and 1100°C in air for one hour.

particle movement and sheath/core strain incompatibility
determine the conditions of core failure, which is
signalled by core and sheath/core interface cracking.
Once limiting core densification is reached, the copper

270

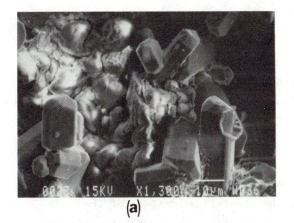

(a)

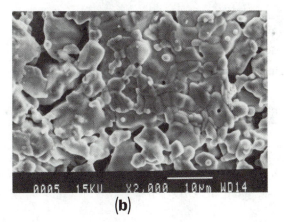

(b)

Figure 12 - Morphology of core materials heated to 1100°C in air: (a) cooled in furnace (3.5°C/min) and (b) air quenched.

sheath is unable to accommodate all the deformations beyond approximately 80% of the original size of the tube and it also cracks.

271

3)   The high-temperature behavior of Y-Ba-Cu-O, at temperatures higher than 560°C, would reduce the success of hot rolling of the powder-packed tubes.

4)   Copper metal powders incorporated into the core material also experience Cu./Y-Ba-Cu-O interface cracking.

5)   The critical current of the wires obtained by cold rolling at optimal conditions is limited by the accentuation of grain boundary resistances, since the final core density achieved by compaction is only a fraction of the theoretical density of the superconductive powder. Interparticle grain boundary resistances also contribute to the low critical currents observed.

Our current investigations include ways to increase compaction density by the incorporation of Bi-based compounds in the core material.

ACKNOWLEDGMENTS

The Colorado Advanced Materials Institute (AMI) and the CSM Engineered Materials Center are gratefully acknowledged for supporting part of this work.

REFERENCES

1. B. Yarar, J. Trefny, F. Schowengerdt, N. Mitra, and G. Pine, Ad. Ceramic Mtls., 2,(3B), 372 (1987).
2. N. Mitra, J. Trefny, M. Young, and B. Yarar, Phys. Rev. B., 36(10), 5581 (1987).
3. N. Mitra, J. Trefny, B. Yarar, G. Pine, Z. Z. Sheng, and A. M. Hermann, Thermoelectric Power of Tl-Ca-Ba-Cu-O, submitted to Phys. Rev., March 10 (1988).
4. R. Roberge, in Proc. NATO ASI on Superconductor Matl. Scie., S. Foner and B. B. Schwartz (eds.), Plenum, New York (1981), pp. 389-453.

5. J. E. Kunzler, E. Buchler, F. S. L. Hus, and
    J. Vernick, Phys. Rev. Lett., 6, 89 (1961).
6. R. W. McCallum, J. D. Verhoeven, M. A. Noack,
    E. D. Gibson, F. Laabs, D. K. Finnemore, and
    A. R. Moodenbaugh, Adv. Ceramic Mtls., 2(3B), 388
    (1987).
7. J. Jin, R. C. Sherwood, R. B. Van Dover,
    T. H. Tiefel, and D. W. Johnson, paper presented at
    MRS Mtg., April 21, Anaheim, California (1987).
8. D. H. Liebenberg, NSF Tokyo Office, Report Memo:
    140, September 30 (1987).
9. L. E. Murr, A. W. Hare, and N. G. Eror, Adv. Matls.
    and Processes, 32(4), 36 (1987).
10. Anon, Logos, 5(3), 13 (1987).
11. M. K. Malik, V. D. Nair, R. V. Raghavan, P. Chaddah,
    P. K. Mishra, G. R. Kumar, and B. A. Dasannacharya,
    Pramana, 29(3), L-321 (1987).
12. D. W. Green (ed.), Perry's Chemical Engineers'
    Handbook, 6th edition, McGraw-Hill, New York, (1984)
    see Chapter 21.
13. B. Yarar and H. Bird, US Pat. Appl. 153208 (1988).
14. E. M. Savitskii, Superconducting Materials, Tnsl.:
    G. D. Archad, Plenum, New York, (1973), p. 400.
15. R. B. Poeppel, B. K. Flandermeyer, J. T. Dusek, and
    I. D. Bloom, in Chemistry of High-Temperature
    Superconductors, D. L. Nelson, M. S. Whittingham,
    and T. F. George (eds.), ACS, Washington, D.C.,
    (1987), pp. 261-265.

SOL-GEL ROUTES TO SUPERCONDUCTING CERAMICS

JAGADISH C. DEBSIKDAR
Idaho National Engineering Laboratory
EG&G Idaho, Inc., Idaho Falls, ID 83415, U.S.A.

Abstract Potential processing advantages of sol-gel
techniques for fabricating ceramic end products are
discussed with special reference to some typical
sol-gel approaches outlined in the literature for
producing high temperature ceramic superconducting
materials. Moreover, the current state of sol-gel
research at the Idaho National Engineering Laboratory
is reported with reference to synthesis of
metallo-organic precursor materials, sol-gel
processing with those precursor materials, and
characterization of high temperature superconducting
materials in the $BaO-Y_2O_3-CuO$ system.

INTRODUCTION

The recent discoveries of high temperature superconducting
materials[1-5] have led to enormous amounts of research
concerning the fundamental aspects of superconducting
materials and the development of commercially significant
end products. The fundamental research efforts are
primarily directed toward the search for new materials with
increasingly higher critical temperature (T_c) and to
understanding the physical principles of superconducting
phenomena in relation to their crystal structures. Appli-
cations-oriented research is concerned with successful
development of enabling technologies for producing superconducting end products. Much of this product development

275

relates to synthesis of chemically homogeneous and uniform particles of the desired crystallographic structure and composition, and the development of processing technology to produce superconducting ceramics in various forms (e.g., films, tape or filament, wires, monoliths) with optimum critical properties relevant to specific applications. This paper critically examines some selected solution chemistry approaches to synthesis of $YBa_2Cu_3O_{7-x}$ superconducting materials, and reports the results of ongoing research using this approach at the Idaho National Engineering Laboratory.

SOLUTION CHEMISTRY APPROACHES TO SYNTHESIS OF $YBa_2Cu_3O_{7-x}$ MATERIALS

The conventional synthesis approach for producing multi-component ceramic oxides, which involves mixing/milling and calcination of oxides and/or carbonates, suffers from the problems of compositional inhomogeneity, uncontrolled particle morphology (i.e., size distribution and shape), particle agglomeration, and impurity pickup during processing steps such as grinding/milling. Because of these common problems, it might be extremely difficult, if not impossible, to fabricate high-purity, high-density superconducting ceramic products of uniform and reproducible microstructures by conventional ceramic fabrication methods. It is well recognized that chemically homogeneous, high purity, multicomponent oxide particles can be synthesized by solution chemistry approaches[6-10] because (1) these approaches allow mixing of the constituent precursor materials at the "molecular level," and (2) no significant grinding/milling is needed to produce the

ceramic particles. Additional advantages relate to the
formation of very fine, spherical, and uniform particles.
Although smaller particle sizes with increased surface free
energy can lead to significantly lower sintering temperatures, particle uniformity contributes to microstructural
uniformity of the final ceramic and thus can ensure optimum
and reproducible product quality. The solution methods of
synthesis can be classified into two broadly different
types: (1) the precipitation/coprecipitation process and
(2) the sol-gel process.

Precipitation/Coprecipitation Process

The process in which separation of insoluble molecular
aggregates takes place, as a result of interaction(s) of
the ions in a metal salt solution, is called precipitation.
The coprecipitation process involves simultaneous precipitation of two or more insoluble species from the solution
of a mixture of metal salts. Precipitation completion is
controlled by the solubility product of the precipitate in
the solvent. Because of this and because the rate of
formation (and eventual separation) of the individual
insoluble species in a multicomponent solution could be
different, chemical inhomogeneity can occur in coprecipitated material. To prevent this, the solution chemistry
must be manipulated so that the individual insoluble
species separate simultaneously from the solvent. These
problems are encountered in coprecipitating the hydrous
oxides of yttrium, barium, and copper from the aqueous
solution of the metal salts when a base, such as ammonium
hydroxide, is used as the precipitant. Besides, copper
forms a soluble complex with ammonium hydroxide. However,
Bunker et al.[11] have overcome these problems by taking a

277

novel approach in which hydroxide-carbonates of yttrium, barium, and copper can be coprecipitated from the aqueous solution using a mixture of tetramethylammonium hydroxide and tetramethylammonium carbonate. Coprecipitation of mixed citrates[12] and oxalates[13] has also been reported for synthesis of $YBa_2Cu_3O_{7-x}$ materials.

Sol-Gel Process

The term sol-gel is broadly used to refer to two methods for forming various glass and ceramic materials in different product forms from solutions. Sols are stable dispersions of ultrafine particles in a suitable solvent. Sols of hydrated oxides can be prepared in three different ways: dispersing extremely fine oxide particles in water, dispersing hydrous oxide precipitate with a small amount of acid(s) (peptization), and synthesis of particles in situ in nonaqueous solvent. Gelation can be described as controlled and oriented flocculation of the sol particles. During the sol-gel transition, the solution viscosity increases gradually at first and, finally, the overall medium is rapidly converted into a quasihomogeneous "solid-like" mass such that no concentration gradient develops. The two methods of sol-gel synthesis are based on two different chemical principles. The method involving preparation of sol either from fine oxide particles or peptization of hydrous oxide precipitate is based on the principles of classical colloid chemistry. In this method, the stability of the sol is controlled by surface charges of the particles, and the sol-gel transition can be affected by destabilizing the surface charges on the colloidal particles. On the other hand, the sol-gel method in which the sol particles are synthesized in situ in the (nonaqueous) solvent is

based on the principles of polymerization chemistry. In
this method, the sol particles are formed by hydrolytic
polycondensation reactions of metallo-organic precursor
compound and localized network formation, and the sol-gel
transition occurs as a result of continued hydrolytic
polycondensation reactions and chemical linkage of the sol
particles. However, each of these methods can have several
variations in solution chemistries. Synthesis of
$YBa_2Cu_3O_{7-x}$ by both these methods has been reported.[14-19]
The method developed by Barboux et al.[14] is similar to
the classical sol-gel process. They mixed hydrous oxide
sol of yttrium (produced from yttrium nitrate by an ion
exchange technique using a commercial anionic resin), an
acidified solution or sol of hydrated barium hydroxide, and
copper acetate solution to produce a precipitate (or
coprecipitate) which turned into a sol after a few hours.
It would appear that free acetic acid added with barium
solution and the acetic acid produced by hydrolysis of
copper acetate acted as the peptizing agent to form a clear
sol. This sol can be used for depositing superconducting
film and for powder preparation as well.

Hirano et al.[15] have reported a chemical
polymerization approach for producing gel-derived films and
fibers. The method involved hydrolysis of a mixed solution
of yttrium isopropoxide, in situ generated barium
ethoxyethanolate, and copper acetylacetonate or copper
ethoxyethanolate in ethoxyethanol solvent with
stoichiometric amounts of water. The material thus
produced showed the Meissner effect after heat treatment at
600°C in an oxidizing atmosphere. Moreover, a partial
hydrolysis of the mixed metallo-organic compounds produced

a sol suitable for depositing superconducting films. Gross
et al.[16] and Kordas et al.[17] have also reported a
metallo-organic approach for producing sol-gel derived
$YBa_2Cu_3O_{7-x}$ superconducting ceramics. The sol-gel
methods reported by Arcangeli et al.[18] and Amarakoon
et al.[19] are somewhat different. Arcangeli et al.[18]
used two different methods for producing $YBa_2Cu_2O_{7-x}$
powders. In one method, the aqueous nitrate solution was
sprayed into a 1:1 mixture of Primene Jmt (a primary amine)
and benzene. In the second method, the nitrate solution in
ethylene glycol was sprayed into triethylamine. The
chemical principles of both these methods relate to
extraction of nitrate ions and simultaneous substitution of
either hydroxyl ion (aqueous approach) or formation of
(perhaps) hydroxyethoxide complexes (nonaqueous approach)
to produce sol-gel powder. Amarakoon et al.[19] used
polyacrylic acid as a gelling agent to produce sol-gel
derived $YBa_2Cu_3O_{7-x}$ powder from an aqueous solution
of metal nitrates. These authors have reported that the
addition of polyacrylic acid at the appropriate solution pH
caused immediate chelation of the cations with the carboxyl
units of the polymer chain, gelling the system. The
material thus produced was first heated at 400 to 450°C
for 3 h and then calcined at 870°C or less to produce the
powder.

Sol-gel synthesis of $YBa_2Cu_3O_{7-x}$ being investigated at
the Idaho National Engineering Laboratory is based on the
chemical polymerization approach. In this paper the
results of our investigations on sol-gel derived
$YBa_2Cu_3O_{7-x}$ and (partially) chlorine substituted
$YBa_2Cu_3O_{7-x}$ are reported. The study on chlorine

substituted $YBa_2Cu_3O_{7-x}$ was included on the basis of
DiSalvo's hypothesis[20] that anions such as chlorine
(also, sulfur and nitrogen) are likely to be good superconducting candidates if metallic phase(s) could be
prepared in this system(s).

EXPERIMENTAL RESULTS

Starting Chemicals

The metallo-organic precursor chemicals used for sol-gel
synthesis of $YBa_2Cu_3O_{7-x}$ gel were: copper amyloxy
acetylacetonate solution in tetrahydrofuran (also containing some amount of amyl alcohol), yttrium amyloxide solution
in amyl alcohol, and a barium alkoxide solution. For synthesizing chlorine-substituted $YBa_2Cu_3O_{7-x}$, yttrium monochlorodiamyloxide solution in amylalcohol was used as the
precursor for both chlorine and Y_2O_3 with a starting
nominal composition corresponding to 2.15 wt% chlorine.
Except for barium alkoxide, which was procured from Dynamit
Nobel Co. (New Jersey), all other chemicals were synthesized in our laboratory by nucleophillic substitution
reactions using metal chlorides. The synthesis procedures
for these metallo-organic compounds appear to be patentable
and, as such, will be published in the future.

Preparation of Gel-Derived Powders

$YBa_2Cu_3O_{7-x}$ gel powder was prepared by the following processing steps. The reactions were carried out in a round-
bottomed, two-neck flask sealed with rubber septa and
purged with nitrogen before transferring the reactants with
a syringe. Step 1: Calculated amounts of barium alkoxide
and yttrium triamyloxide solutions were added, in that
sequence, to the solution of copper amyloxy

acetylacetonate, stirred for ~1 h, and then hydrolyzed at the ambient temperature with 2 moles deionized water per mole of alkoxide mixture to form a blue gel. Step 2: The gel was vacuum dried in a glove box and then heated under vacuum at 140°C overnight to remove free alcohol and some of the residual organics. The powdery material thus obtained was light black in color. Step 3: The dried material was ground in an agate mill.

The processing steps for preparing chlorine substituted $YBa_2Cu_3O_{7-x}$ gel powder were: Step 1: Barium alkoxide was added to copper amyloxy acetylacetonate to form a clear solution. An addition of yttrium monochlorodiamyloxide to this solution produced a precipitate. At this stage, most of the solvents (tetrahydrofuran and amylalcohol) were pumped off, and methoxyethoxyethanol was added to form a clear solution, which, on hydrolysis with 5 moles deionized water, produced a bluish green gel at 60°C overnight. Steps 2 and 3 were the same as those for preparing $YBa_2Cu_3O_{7-x}$ gel powder.

Characterization of Gel-Derived Powders

Thermal gravimetric and differential thermal analysis (TGA and DTA) curves for $YBa_2Cu_3O_{7-x}$ and chlorine-substituted $YBa_2Cu_3O_{7-x}$ gel powders obtained at the heating rate of 5°C/min in free-flowing oxygen are shown in Figures 1 and 2, respectively. The weight losses of $YBa_2Cu_3O_{7-x}$ and chlorine substituted $YBa_2Cu_3O_{7-x}$ gel powders were ~50 and 37%, respectively (Figure 1). These losses were essentially complete at 350°C or less, and were associated with two exotherms each of the DTA curves in this temperature range (Figure 2). Therefore, the losses could be attributed to oxidative removal of the residual organics

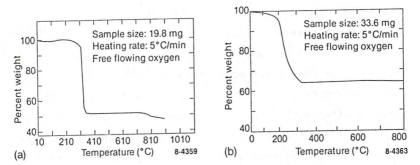

FIGURE 1. TGA of gel powders: (a) $YBa_2Cu_3O_{7-x}$, and
(b) chlorine substituted $YBa_2Cu_3O_{7-x}$.

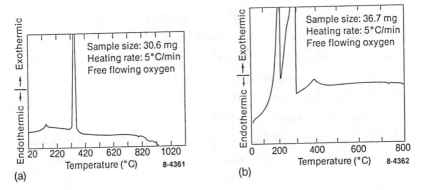

FIGURE 2. DTA of gel powders: (a) $YBa_2Cu_3O_{7-x}$,
and (b) chlorine substituted $YBa_2Cu_3O_{7-x}$.

and presumably to evolution of crystalline phase(s) as
well. The additional exotherm observed in the chlorine
substituted $YBa_2Cu_3O_{7-x}$ gel at 400°C (Figure 2b) was not
associated with weight loss (Figure 1b) and, moreover, such
an exotherm was not observed with $YBa_2Cu_3O_{7-x}$ gel
(Figure 1a). Therefore, this peak might represent
evolution of an oxychloride phase. Both gel powders showed
an extremely small endotherm in the neighborhood of
750°C. It is conceivable that a fraction of CO_2 formed
during the oxidative removal of organics reacted with

283

either free barium oxide, barium hydroxide, or barium
containing binary or ternary compound to form $BaCO_3$
(Ref. 21), which decomposed at ~750°C. The endotherm
at ~850°C or less observed with these powders might
represent some melting phenomenon. The crystallinity of these
powders after 850°C/2 h heat treatment (in free-flowing
oxygen) is shown in Figure 3. Both of these powders showed
orthorhombic phase and, in addition, several extra peaks.
In the case of $YBa_2Cu_3O_{7-x}$ material, the extra peaks
were compared with standard reference peaks of the
individual and binary oxides of yttrium, barium, and copper
($Cu_2Y_2O_5$, $Ba_4Y_2O_7$, $BaCuO_2$, $Ba_3Y_4O_9$ and
$Ba_2Y_2O_5$), and $BaCO_3$. The extra peaks, however,
could not be satisfactorily correlated to these compounds.
Therefore, we presume that these additional peaks might
represent an unknown compound or a complex mixture of some
of the above compounds. The additional diffraction peaks
of chlorine substituted $YBa_2Cu_3O_{7-x}$ might be related

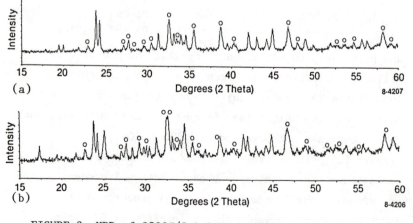

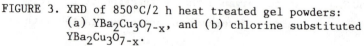

FIGURE 3. XRD of 850°C/2 h heat treated gel powders:
(a) $YBa_2Cu_3O_{7-x}$, and (b) chlorine substituted
$YBa_2Cu_3O_{7-x}$.

284

to the presence of one (or more) barium copper oxychloride phase(s) which can be seen later in EDX spectrometric results.

Fabrication and Characterization of Ceramics

For ceramic fabrication, the gel powders were freed from organics by calcination at 500°C in a tube furnace in free-flowing oxygen (flow rate $\sim$200 mL/min). The morphology of organic-free $YBa_2Cu_3O_{7-x}$ powder was examined by scanning electron microscopy (AMR Model 1200) after ultrasonically dispersing it in anhydrous alcohol. The powder was highly agglomerated with agglomerate size range 4 to 5 μm in diameter (Figure 4). However, these agglomerates were composed of spherical particles 0.4 to 0.5 μm in diameter. This result indicates that it is feasible to produce submicron size spherical particles of

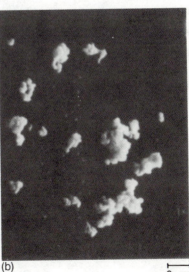
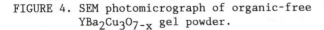

(a)   10 μm (b)   2 μm

FIGURE 4. SEM photomicrograph of organic-free
$YBa_2Cu_3O_{7-x}$ gel powder.

$YBa_2Cu_3O_{7-x}$ by the sol-gel process. The organic-free $YBa_2Cu_3O_{7-x}$ and chlorine substituted $YBa_2Cu_3O_{7-x}$ gel powders were uniaxially pressed at $\sim$345 MPa to produce $\sim$1.27 cm dia x 0.63 cm thick "green" pellets without adding any organic binder. These pellets were sintered at 950°C for 2 h and then annealed at 500°C by a sintering schedule which involved the following ramps: heating to 950°C at 100°C/h , holding at 950°C for 2 h, cooling to 500°C at 50°C/h, holding at 500°C for 10 h, and cooling to room temperature at 50°C/h. A 950°C sintering temperature was used so that the microstructures and chemical homogeneity of sol-gel derived ceramics could be compared with those of the conventionally derived ceramics which are generally sintered at this temperature, although with a much longer hold (8 to 24 h or more). The XRD of the $YBa_2Cu_3O_{7-x}$ ceramic pellet (Figure 5) showed that the material was single-phase orthorhombic in crystallographic form. On the other hand, the XRD of chlorine-substituted material still showed additional peaks as observed in 850°C heat treated powder (Figure 3b). These ceramic pellets were characterized in terms of resistance versus temperature in an exchange gas controlled liquid nitrogen dewar by the AC four terminal method using a 15% Pb, 5% Ag, and 80% In

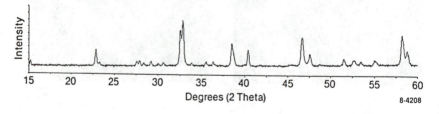

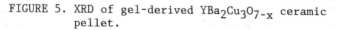

FIGURE 5. XRD of gel-derived $YBa_2Cu_3O_{7-x}$ ceramic pellet.

solder. Microstructures of the pellets were examined by
SEM and chemical homogeneity by energy dispersive x-ray
(EDX) spectrometry.

Both $YBa_2Cu_3O_{7-x}$ and chlorine-substituted
$YBa_2Cu_3O_{7-x}$ ceramics showed zero resistance at 89 K
(Figure 6) with the superconductivity onset temperatures 91
and 92 K for $YBa_2Cu_3O_{7-x}$ and chlorine-substituted
$YBa_2Cu_3O_{7-x}$ ceramic, respectively. However, the
resistance versus temperature curve of chlorine-substituted
ceramic showed a tail end which could be related to
chemical inhomogeneity of the material. The
microstructures of both $YBa_2Cu_3O_{7-x}$ and chlorine
substituted $YBa_2Cu_3O_{7-x}$ (Figure 7) consisted of
mostly platelike crystals of various sizes and nonuniform
pore morphology. The presence of intragranular porosity
was also observed (Figure 7b). The abnormal grain growth
observed in these specimens presumably indicates that
viscous sintering occurred at 950°. The EDX spectrometry
of $YBa_2Cu_3O_{7-x}$ showed that, except for the crystals
that showed contamination of sodium and/or silicon
impurities, the spectra of other crystals were similar

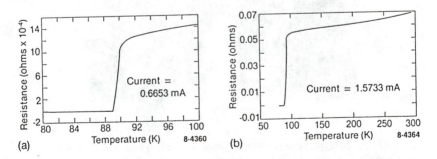

FIGURE 6. Resistance vs. temperature of gel-derived
ceramic pellets: (a) $YBa_2Cu_3O_{7-x}$, and
(b) chlorine-substituted $YBa_2Cu_3O_{7-x}$.

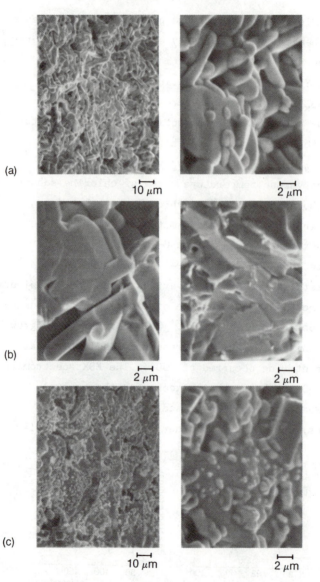

FIGURE 7. SEM photomicrograph of gel-derived ceramics:
(a) as-fired surface of $YBa_2Cu_3O_{7-x}$,
(b) fracture surface of $YBa_2Cu_3O_{7-x}$, and
(c) as-fired surface of chlorine-substituted
$YBa_2Cu_3O_{7-x}$.

288

within experimental error (Figure 8). The crystals of chlorine substituted ceramic, however, showed a high degree of compositional inhomogeneity (Figure 9). Most of the chlorine was associated as barium-copper oxychloride, and the yttrium-rich phase contained only traces of chlorine even though chlorine was introduced in this material as yttrium monochlorodiamyloxide. Moreover, this material contained copper oxide crystals.

The compositional homogeneity of gel-derived ceramics was also compared with a conventionally prepared single phase $YBa_2Cu_3O_{7-x}$ (prepared in our laboratory) with T_c close to 90 K. The SEM microstructure and EDX

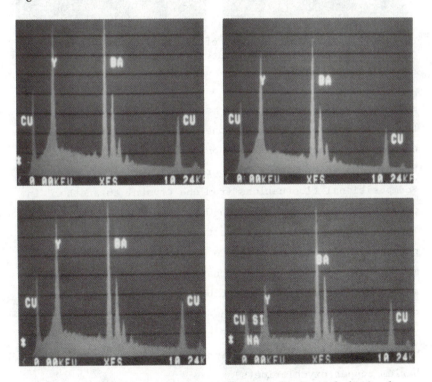

FIGURE 8. EDX spectrometry of crystals in gel-derived $YBa_2Cu_3O_{7-x}$ ceramic.

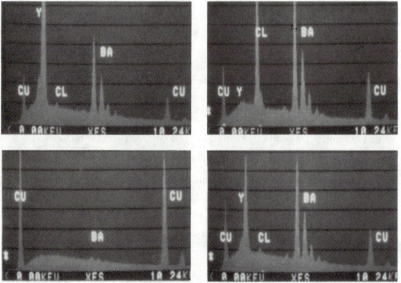

FIGURE 9. EDX spectrometry of crystals in gel-derived chlorine–substituted $YBa_2Cu_3O_{7-x}$ ceramic.

spectrometry of this material is shown in Figure 10. Of the three crystals examined, one crystal was found to be essentially copper oxide with only traces of yttrium and barium. Gasgnier et al.[22] reported even wider "compositional fluctuations from one crystal to another" in conventionally produced $YBa_2Cu_3O_{7-x}$ with T_c ∿90 K. These results indicate that (1) the sol-gel process offers the potential to produce compositionally more homogeneous $YBa_2Cu_3O_{7-x}$ than the conventional approach, and (2) it might not be possible to synthesize chemically homogeneous chlorine–substituted $YBa_2Cu_3O_{7-x}$ ceramic because of the preferential reaction of chlorine with barium and copper to form barium-copper oxychloride(s).

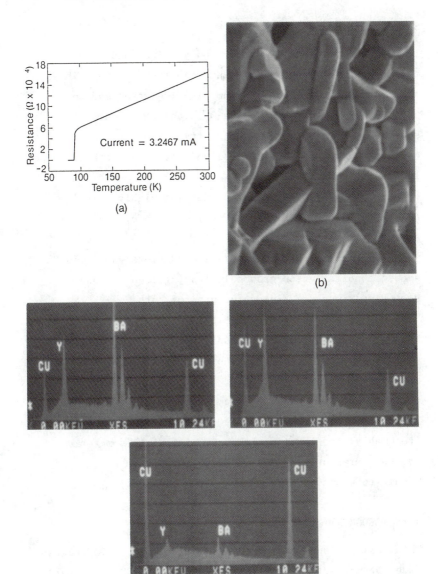

(a)

(b)

(c)

FIGURE 10. Characteristics of conventionally produced
$YBa_2Cu_3O_{7-x}$: (a) resistance vs.
temperature, (b) SEM microstructure, and
(c) EDX spectrometry of crystals.

SUMMARY AND CONCLUSIONS

1. Alkoxy compounds of copper and yttrium, and chloroalkoxy compound of yttrium were synthesized. These alkoxides and commercial barium alkoxide were used to produce sol-gel derived $YBa_2Cu_3O_{7-x}$ and chlorine substituted $YBa_2Cu_3O_{7-x}$ (with starting composition corresponding to 2.15 wt% chlorine) ceramics, both of which showed T_c around 90 K.

2. The $YBa_2Cu_3O_{7-x}$ gel powder was highly agglomerated but consisted of submicroscopic, spherical, and uniform particles.

3. Sol-gel derived material when sintered at 950°C showed abnormal grain growth indicating liquid phase sintering process. Microstructures of both $YBa_2Cu_3O_{7-x}$ and chlorine substituted $YBa_2Cu_3O_{7-x}$ ceramics were nonuniform.

4. Compositional homogeneity of sol-gel derived $YBa_2Cu_3O_{7-x}$ ceramic was better than conventionally derived ceramic. However, the chlorine substituted $YBa_2Cu_3O_{7-x}$ ceramic was highly inhomogeneous. Presumably, the chemical inhomogeneity of this material was due to preferential reaction of chlorine with barium and copper to produce barium-copper oxychloride(s).

ACKNOWLEDGMENT

This work was supported by DOE Fossil Energy AR&TD program under Department of Energy contract No. DE-AC07-76ID01570. The author thanks E. Samsel for synthesizing the alkoxy compounds, K. Telschow for measuring T_cs, and D. Miley for the work of scanning electron microscopy.

SOL-GEL ROUTES TO SUPERCONDUCTING CERAMICS

REFERENCES

1. J. G. Bednorz and K. A. Müller, Z. Phys., B 64, 189 (1986).
2. M. K. Wu et al., Phys. Rev. Lett, 58, 908 (1987).
3. R. J. Cava et al., Phys. Rev. Lett, 58, 408 (1987).
4. U. Chowdhry et al., Presentation at the World Congress on Superconductivity, Houston, Texas, Feb. 22-24, 1988.
5. Z. Z. Sheng et al., ibid.
6. J. M. Fletcher et al., Chem. and Ind., 48, (Jan. 1968).
7. C. J. Hardy in Special Ceramics, 4, edited by P. Popper (The British Ceramic Soc. Assoc., 1968), p. 85.
8. J. L. Woodhead et al., Brit. Pat., 1,266,494 (March 1968).
9. H. Dislich, Angew. Chem (Int. Edn. Engl.) 10, 363 (1971).
10. M. E. A. Hermans, Powder Metallurgy International, 5(3), 137 (1973).
11. B. C. Bunker et al., to be published in High Temp. Superconducting Materials: Preparation, Properties and Processing, edited by W. E. Hatfield and J. H. Miller, Jr. (Marcel Dekker, Inc., New York, 1988).
12. C. T. Chu et al., J. Am. Ceram. Soc., 70(12) C-375 (1987).
13. A. R. Moodenbaugh et al., "Extended Abstract of the Proceedings of Symposium S," 1987 Spring Meeting of the Materials Research Society, April 23-24, 1987, edited by D. U. Gubser and M. Schluter, p. 101.
14. P. Barboux et al., Bell Communication Research, Personal Communication, to be published.
15. S. I. Hirano, et al., Presentation at the International Symposium on "Ceramic Substrate and Packages", Denver, October 18-21, 1987.
16. M. Gross et al., Materials and Processing Report, Jan. 1988 AT&T Bell Laboratories, Murry Hill, NJ.
17. G. Kordas et al., Materials Letters, 5, 11 and 12, 417 (1987).
18. G. Arcangeli et al., Comb-Bapu-Sapo Fuel Cycle Dept., Roma, Poster Presentation at the World Congress on Superconductivity, Houston, Texas, February 22, 1988.
19. V. R. W. Amarakoon et al., Advanced Ceramic Materials, 2, 3B (July 1987).
20. F. J. DiSalvo, in Chemistry of High-Temperature Superconductors, edited by D. L. Nelson, M. S. Whittingham, and T. F. George (American Chemical Society, Washington, D.C.), Chapter 5, p. 49, August 30-September 4, 1987.

21. J. C. Debsikdar, <u>Synthesis and Characterization of Gel Derived Barium Hexaaluminate</u>, to be published.
22. M. Gasgnier et al., <u>J. Appl. Phys</u>, 4935 (December 15, 1987).

SYNTHESIS AND PROCESSING STUDY OF $Ba_2YCu_3O_{7-x}$:
IMPROVED METHODS AND PROCESSING-DEPENDENT ALIGNMENT

ALOYSIUS F. HEPP and JAMES R. GAIER
National Aeronautics and Space Administration
Lewis Research Center, Cleveland, OH 44135

Abstract   Solid-state syntheses using BaO_2 and $BaCO_3$
are compared. BaO_2 produces more dense, homogeneous,
and metallic material. C-axis alignment resulted
from firing loose powder at temperatures > 925°C.

INTRODUCTION

Bulk applications of ceramic superconductors require
material free from flaws which limit critical current
density.[1] The presence of intergrain $BaCO_3$[2] in
$Ba_2YCu_3O_{7-x}$ points to the need for its exclusion. This
motivates the search for improved synthetic procedures.[3]
We compare syntheses of $Ba_2YCu_3O_{7-x}$, discuss improved
synthetic methods and present data on aligned material.

EXPERIMENTAL

CuO, Y_2O_3, and $BaCO_3$ (all >99.99%) were obtained from
Cerac; BaO_2 (99%) was obtained from J.T. Baker. Powders or
green pellets were fired at 890°C, 925°C, 950°C, or 975°C
under flowing O_2 or N_2. At 16 hour intervals, samples
were reground, weighed, and analyzed by x-ray. T_c and A.C.
susceptibility methods have been described previously.[4]

RESULTS AND DISCUSSION

Denser (85% versus 60 % theoretical density) and more
homogeneous material was prepared using BaO_2. Figure 1
shows a comparison of EDXS data of two samples which were
fired and processed several times. This result typifies
our experience: BaO_2 use improves material quality.[4]

SYNTHESIZED WITH BaCO$_3$ SYNTHESIZED WITH BaO$_2$

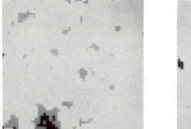 

(A) VARIATION IN Ba PER UNIT Cu.

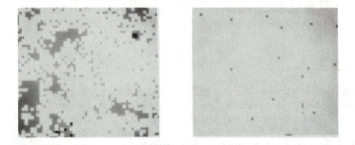

(B) VARIATION IN Cu PER UNIT Y.

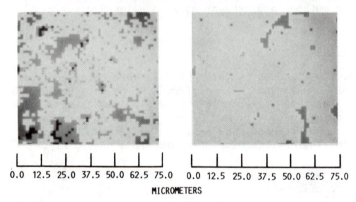

0.0 12.5 25.0 37.5 50.0 62.5 75.0   0.0 12.5 25.0 37.5 50.0 62.5 75.0

MICROMETERS

(C) VARIATION IN Ba PER UNIT Y.

FIGURE 1. Difference Energy Dispersive X-ray
Spectroscopy (EDXS) images show the variation in
composition across the sample. Dark areas indicate
an excess of one of the elements with respect to
the other.

IMPROVED METHODS AND PROCESSING-DEPENDENT ALIGNMENT

Lattice constants in all samples were within Γ .3% (a = 3.83 Å, b = 3.89 Å, and c = 11.69 Å). Oxygen stoichiometries were calculated using results of Wong-Ng[5] and Tarascon[6] and agree to within 5%. Samples made with $BaCO_3$ increased in oxygen stoichiometry from 6.6 to 6.7 to 6.8 from the first to the third firing in both the muffle furnace and flowing oxygen. Reactions run under flowing N_2 using $BaCO_3$ show that the oxygen stoichiometry remains at 6.5. This is consistent with the stoichiometry of starting materials. Complete calcination of $BaCO_3$ in the absence of oxygen results in 2 molecules of CO_2 and 6.5 oxygens indicating the need for external oxygen.

Material made with BaO_2 remained constant at 6.8 oxygens. This result agrees with weight loss data: a 4% weight loss from unreacted powders with BaO_2 produces a $Ba_2YCu_3O_{6.8}$ stoichiometry. Weight losses at our level of accuracy (1 %) are identical, whether the reactions are done in flowing N_2 or flowing O_2, consistent with the stoichiometry and chemistry of the starting materials. The initial stoichiometry of starting materials is $Ba_2YCu_3O_{8.5}$. The reaction could proceed without external oxygen to produce superconducting material. $Ba_2PrCu_3O_{6.95}$ has been made in vacuo using BaO_2.[7] Reactions run in flowing O_2 and N_2 produce single-phase crystalline material. The most noticeable difference in reaction products is that single phase material is produced more rapidly at 950°C.

We found reactions in flowing N_2 to be very sensitive to the N_2 flow rate and the geometry of the reaction vessel. When we increased our N_2 flow rate to 9 liters per minute in a flat Al_2O_3 crucible, we essentially reproduced a TGA reduction and produced tetragonal material with an oxygen stoichiometry close to 6, particularly at higher temperatures. Our reported reaction conditions model much more closely a sealed, static atmosphere or vacuum[7]; O_2 evolved from BaO_2 decomposition could interact with solids.

The A.C. susceptibility of ground pellets was 50% (corrected to 80% theoretical density) at 68K for samples prepared with $BaCO_3$ or BaO_2, typical of literature results.[3] T_c for samples prepared with BaO_2 or $BaCO_3$ is similiar (90K) with a sharper transition for samples prepared with BaO_2 (ΔT_c = 2K versus ΔT_c = 4K). BaO_2 produced samples had greater metallic character, as determined by the slope of resistivity versus temperature, than similiarly produced material using $BaCO_3$. BaO_2 produced samples also had lower room temperature resistivity by a factor of three. This result is consistent with less resistive grain boundaries.

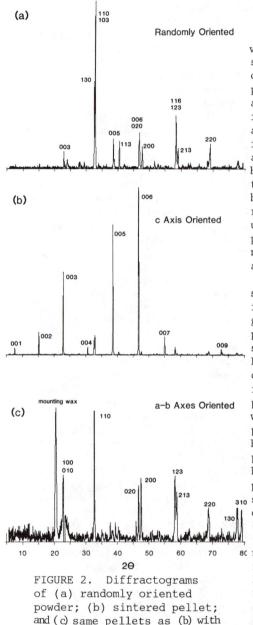

FIGURE 2. Diffractograms of (a) randomly oriented powder; (b) sintered pellet; and (c) same pellets as (b) with reflections taken edge on.

XPS measurements were carried out on in-situ fractured pellets of a number of samples prepared as described above. A 25% reduction in intergrain $BaCO_3$ was achieved; previously, intergrain $BaCO_3$ was also not eliminated.[2] However, BaO_2 use allows total inert atmosphere handling. Given the reactivity of CO_2 with unreacted barium complexes, this appears necessary for elimination of $BaCO_3$.

During our precursor study, we observed interesting diffracto-grams of reaction products; this is shown by Figure 2. Figure 2a shows a diffractogram of an isotropic powder; this pattern was obtained when pressed green pellets were fired. However, when loose powders were fired, at higher temperatures, then pressed into pellets and sintered, this produced orientationally enhanced (00ℓ) x-ray intensities, (Figure 2b); x-ray data normal to the pellet face showed the hk0 lines with enhanced intensities (Figure 2c). Grains produced from loose powder reaction were

platelet-like with aspect ratios greater than 3:1; when
compressed uniaxially, the platelets oriented perpendicular
to stress along the c axis. We found that faceting was
enhanced by use of $BaCO_3$, due to liquid phase sintering of
previously formed platelets. Our results support crystal
growth kinetics as the driving force behind platelet
growth. A model to describe the formation of platelet-
shaped grains is shown in Figure 3. Faceted growth occurs
along the c axis, since this slow-growth plane is limited by
surface nucleation. As the a-b plane is the fast-growth
plane, mass transport limits due to firing of loose powder
would lead to unfaceted growth.

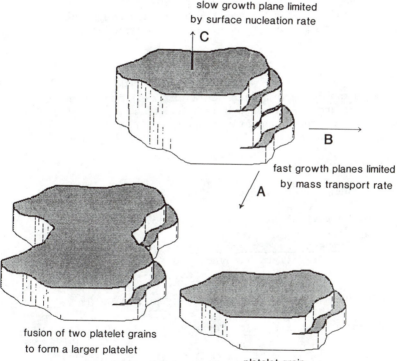

FIGURE 3. Model for occurance of faceted grain
growth. Top shows faceting along slow growth
plane. Bottom shows sintering of platelets to form
a larger platelet (see text).

CONCLUSIONS

Superconducting material made using BaO_2 is of higher quality because of the reactivity of BaO_2. The stability of $BaCO_3$ makes complete calcination difficult, resulting in non-uniform material. BaO_2 use increases reaction rates and produces higher-quality material more routinely. We have shown in our reaction data, x-ray data, and critical temperature measurements that superconducting $Ba_2YCu_3O_{7-x}$ can be produced in inert atmosphere using BaO_2. The reactivity afforded by highly oxidized compounds leads to improved material. It has recently been reported that KO_2 has been used to produce $Ba_{0.6}K_{0.4}Bi_{0.3}O_x$, a non-cuprate superconductor with a T_c of 30K.[8] Indeed, use of highly oxidizing species may result in new compounds and superior processes not possible with traditional reagents.

ACKNOWLEDGEMENTS

We acknowledge Mr. Ralph Garlick (XRD) for technical assistance and Drs. Sheila Bailey and Geoff Landis for collaboration on alignment phenomena.

REFERENCES

1. R.A. Camps et al., Nature 329, 229 (1987).

2. A.G. Schrott et al., subm. to J. Vac. Sci. Technol.

3. High-Temperature Superconductors, edited by M. Brodsky, H. Tuller, R. Dynes and K. Kitazawa (MRS Proceedings 99, Mater. Res. Soc., Pittsburgh, PA, 1988).

4. A.F. Hepp et al., in reference 3, pp. 615-618.

5. W. Wong-Ng et al., Adv. Ceram. Mat., 2, 565 (1987).

6. J.M. Tarascon et al., in Chemistry of High-Temperature Superconductors, edited by D.L. Nelson, M.S. Whittingham, and T.F. George (ACS Symp. Ser. 351, Amer. Chem. Soc., Washington D.C., 1987). pp. 198-210.

7. J. Amador et al., in reference 3, pp. 249-252.

8. R. Cava et al., Nature, in press.

THE MYTH OF THE POSSIBILITY OF HIGH CRITICAL CURRENT
IN MICROCRYSTALLINE OXIDE WIRES: THE FAILURE OF
THREE DIMENSIONAL PROCESSING TECHNIQUES IN LOWER
DIMENSIONS

MICHAEL L. NORTON
Department of Chemistry, University of Georgia, Athens,
Georgia, 30602, U.S.A.

Abstract   Theoretical   and   experimental   evidence
is   provided   to   support   the   contention   that
implementation of traditional ceramic and metallurgical
methods   of   handling   materials   cannot   succeed   in
preparing   high   critical   current   density   microcables
from   the   low-dimensional   oxides.   By   comparing   and
contrasting   the   new   oxides   with   traditional   ceramic
and alloy materials, the mandate for crystallographic
organization   is   made   obvious.   Methods   of
implementation of such organization of low-dimensional
perovskite related materials are described.

INTRODUCTION

This   paper   is   organized   into   three   sections,   one
explaining the failure, in principle, of low-dimensional
materials in microcrystalline form ever to perform within
an order of magnitude of required J_c's, a second part,
in   which   literature   examples   of   pursuits   of   mythical
high-J_c polycrystalline "wires" are shown, in principle,
and   in   fact,   to   have   intrinsic   deficiencies.   The third
portion   indicates   approaches   lacking   these   intrinsic
deficiencies.

LOW DIMENSIONAL FUNDAMENTALS

Although   this   paper   emphasizes   the   compound $YBa_2Cu_3O_7$
(Y123),   the   other   high T_c oxides   (Bi   and   Tl)   resemble
Y123.

These systems are layered, as opposed to the pseudoanisotropic low temperature (e.g. A15) superconductors. Aside from a possible intrinsic semiconductivity along the c-axis,[1] there is the possibility of insulating stacking faults.[2] A more formidable problem is the co-existence of insulating phases, a problem intrinsic to bulk oxide preparations, but absent in the reduced phase (e.g. A15) alloy materials. The presence of an EPR signal is one indication of impurity phase contamination.[3] These factors will require percolation, reducing the current capacity of these "wires". Grain decoupling studies[4] indicate that the Josephson junction connections at grain boundaries serve to degrade J_c in the presence of low fields, even when the grains themselves are capable of supporting high critical currents.

METHODS OF THE MYTH

The only hazard to the approaches in this section is that they may be considered as viable approaches to the manufacture of high J_c wire.

STANDARD CERAMIC PROCESSING

In this technique, raw or processed materials are usually mixed with organic binders, dried to a green state, and pass through steps allowing outgassing of the binder decomposition products. This method must fail due to: (1) presence of grain boundaries, (2) lack of grain alignment and (3) decoration of grain boundaries by the decomposition products associated with the binder. Evidence for this failure is well documented, with J_c's of 90 to 725 A/cm^2 (77K) reported.[5,6]

MYTHS OF HIGH CRITICAL CURRENTS

WIRE DRAWING

The oxide superconductor can be placed into tubes of a malleable metal and be swaged. Although this may result in some ordering of the individual crystallites due to the anisotropic growth habit of the layered materials, the grain boundaries and any decorations of them will persist, as well as there being the possibility that the grain boundaries will be further decorated by silver atoms[7]. The low J_c's (eg. 640 A/cm^2)[7] reported for this type of wire reflects the intrinsic macroscopic defects in the starting material, and should preclude further pursuit of this approach.

ALLOY OXIDATION

It has been suggested that a dense superconducting oxide could be prepared through the oxidation of an alloy of the metallic elements of the superconductor.[8,9] This approach must fail in principle, since these are not interstitial alloys. Evidence for the flaking and spalling normally observed to result from this process has been reported.[8] The change in volume accompanying oxidation will disrupt any orderly growth, leading to a mechanically weak assembly of randomly oriented superconducting grains. The J_c reported, 20 A/cm^2 (4.2K),[9] reflects the intrinsic failure of this approach.

BETTER METHODS: OPTIMUM SOLUTIONS TO DIFFICULT PROBLEMS

From the above discussion, it becomes clear that wire approaching single crystallinity represents the only viable option for high current applications. Methods for quasi-single crystalline wire production are reviewed here.

AMORPHOUS MATERIALS

From the first reported experiments by Jin et al.[10], to their most recent report of $J_c = 1.7 \times 10^4$ A/cm^2 at 77K,[11] this method has been shown to be far superior to the other polycrystalline approaches. It is not clear that attempts have been made to accomplish an amorphous to single crystalline transition, as is used to produce single crystalline thin films[12] rather than an amorphous to polycrystalline transition during the annealing stage. Recent reports of the solubility of oxygen in the melt of Y123 indicate that a melt approaching the O_7 ($YBa_2Cu_3O_7$) stoichiometry can be achieved.[13] A quenched melt of this composition would not necessarily undergo a large volume change at the amorphous/crystalline transition, a prerequisite of single crystal formation. Key to this approach will be the use of autoepitaxy of a travelling crystallization front running the length of the fiber.

THICK FILM APPROACHES

The same methods which have been used to create thin films could be applied to the preparation of thick films, perhaps on wire (fiber) shaped substrates, such as MgO. Sputtering and chemical vapor deposition techniques are already proven techniques as fast methods of single crystal film growth. It may even be possible to grow such single crystal films in a detachable, free standing film format.

FIBRE STRUCTURE DESIGN AND GROWTH

More challenging than the previous methods is the possibility of preparing the oxide superconductors as "asbestiform" minerals. These minerals are well known as multi-layered oxides, with structures related to those of the oxide superconductors (Ln, Bi, and Tl). Although the chemistry of these materials differ greatly, it is possible that the structural resemblances are more important. The presence of, and capacity for, planar defects is correlated with the propensity for growth in a fiber morphology.[14] Both the Bi and Tl system appear to be capable of supporting intergrowth by two very similar phases.[15,16] Although we are apparently far from having the option of choosing the structure for a superconductor while keeping the properties optimized, this aspect may be very important in determining the technological viability of these oxide materials.

CONCLUSION

Of the processing techniques available, the undue emphasis on impractical, traditional approaches is reflected in the technical literature. An understanding of the basic differences between these quasi- 2 and traditional 3-dimensional materials will lead to use of processing technique which deal with these materials in a manner fundamentally different than those traditionally employed with traditional, "isotropic" materials.

ACKNOWLEDGEMENT

This work is supported by a grant from the National Science Foundation, CHE 8600224.

REFERENCES

(1) M. Suzuki, M. Y. Enomoto and T. Murakami, Jpn. J. Appl. Phys., 26, no. 12, L2052 (1987).

(2) A. Ourmazd, J. A. Rentschler, J. C. H. Spence, M. O'Keeffe, R. J. Graham, D. W. Johnson, Jr. and W. W. Rhodes, Nature, 327, no. 6120, 308

M. L. NORTON

(1987).

(3)   J. A. O. de Aguiar, A. A. Menovsky, J. vanden Berg and H. B. Brown, J. Phys. C: S. S. Phys., 21, L237 (1988).

(4)   J. F. Kwak, E. L. Venturini, D. S. Ginley and W. Fu in Novel Superconductivity (S. A. Wolf and V. Z. Kresin, Ed.), Plenum Press, New York, 1987, pp. 983-991.

(5)   M. K. Malik, V. D. Nair, R. V. Raghavan, P. Chaddah, P. K. Mishra, G. R. Kumar and B. A. Dasannacharya, Pramana-J. Phys., 29, no. 3, L321 (1987).

(6)   Y. Yamada, N. Fukushima, S. Nakayama, H. Yoshino and S. Murase, Jpn. J. Appl. Phys., 26, no. 5, L865 (1987).

(7)   O. Kohno, Y. Ikeno, N. Sadakata, M. Sugimoto and M. Nakagawa, Physica, 148B, 429 (1987).

(8)   G. Y. Yurek, J. B. V. Van der Sande, W. X. Wang, D. A. Rudman, Y. Zhang and M. M. Matthiesen, Metallurgical Trans. A, 18A, 1813 (1987).

(9)   K. Matsuzaki, A. Inoue, H. Kimura and T. Masumoto, Jpn. J. Appl. Phys., 26, no. 10, L1610 (1987).

(10)   S. Jin, T. H. Tiefel, R. C. Sherwood, G. W. Kammlott and S. M. Zahurak, Appl. Phys. Lett., 51, no. 12, 943 (1987).

(11)   R. Pool, Science, 240, 25 (1988).

(12)   J. Kwok, T. C. Hsieh, M. Hong, R. M. Fleming, S. H. Liou, B. A. Davidson and L. C. Feldman, Conference Proceedings, Materials Research Society Fall Meeting, Boston, (preprint) (1987).

(13)   J. Karpinski and E. Kaldis, Nature, 331, 242 (1988).

(14)   G. N. Subbanna, T. R. N. Kutty and G. V. Anantha Iyer, American Mineralogist, 71, 1198 (1986).

(15)   D. R. Veblen, P. J. Heaney, R. J. Angel, L. W. Finger, R. M. Hazen, C. T. Prewitt, N. L. Ross, C. W. Chu, P. H. Hor and R. L. Meng, Nature, 332, 334 (1988).

(16)   I. Peterson, Science News, 133, 148 (1988).

PROCESSING OF FLEXIBLE HIGH-T$_c$ SUPERCONDUCTING WIRES

B. I. LEE and V. MODI
Ceramic Engineering, Clemson University, Clemson, SC
29634, USA

Abstract. Wires superconducting at temperatures above 77 K
are produced by using $YBa_2Cu_3O_7$ materials. Flexibility was
obtained by support from prefabricated fibers or a metallic
coating on the extruded $YBa_2Cu_3O_7$ wires. The microstructure,
the T_c and the critical current densities of the wires were
determined. Processing variables and steps are described.

THE ROLE OF THE METALLIC ORBITAL AND OF CREST AND TROUGH SUPERCONDUCTION IN HIGH-TEMPERATURE SUPERCONDUCTORS

LINUS PAULING
Linus Pauling Institute of Science & Medicine
Palo Alto, California

Abstract   In the resonating-covalent-bond
theory of the structure of metals, about 44%
of the metal atoms are electrically neutral,
and 28% with a positive charge, 28% with a
negative charge. In small volumes the
resultant electric charge is 0, but,
because of statistical fluctuations large,
clusters with an extra positive charge or
negative charge can form. These clusters,
electric-charge fluxons, interact with one
another in such a way as to sustain an
electric current at low temperatures.
Electron-phonon interactions tend to quench
the current. High-temperature superconductivity results when the superconductor
consists of both hypoelectronic and hyperelectronic metals. A way of calculating the
strength of the electron-phonon interaction
for a crest-trough superconductor is
discussed.

In 1938 I formulated a resonating-covalent-bond
theory of metals.[1] I was impressed by the fact
that chromium, manganese, iron, cobalt, and nickel
are far denser, harder, and stronger than
potassium and calcium, and I concluded that in
these transition metals there are resonating
covalent bonds corresponding to valence approximately 6, rather than the much smaller value of
the valence assigned by Slater, Mott, and others.

This assignment involved use of the 4s and 4p orbitals, as well as the 3d orbitals. For iron, in metallic iron, for example, I assigned the covalence 5.78, with 2.22 electrons serving as ferromagnetic electrons. I pointed out that for cobalt and nickel a total of 8.28 orbitals were used for bonding, for ferromagnetic electrons, or for unshared electron pairs, leaving 0.72 orbital per atom unused. Later, in 1948 and 1949 I recognized that the 0.72 orbital per atom, that seemed not to be used, was in fact available for occupancy by an additional electron, in order to permit unsynchronized resonance.[2,3] I named this the metallic orbital, characteristic of metals. The value 0.72 is attributed to there being 28% M^+, 44% M^0, and 28% M^-. The metallic orbital is required only for M^+ and M^0, in order that they be able to accept an additional bonding electron.

A statistical theory of unsynchronized resonance of covalent bonds in metals was mentioned in my 1949 paper, and then developed further in 1984 and 1985.[4,5] It was then shown[6] that it leads to exactly the experimental value, 0.72 orbital per atom, for the alloy between nickel and copper that is the foot of the ferromagnetic moment curve.

In 1968 I discussed a mechanism for the interaction of an electric current in metals with phonons.[7] This discussion permitted metals to be divided into three classes, hypoelectronic metals, buffer metals, and hyperelectronic metals.

THE ROLE OF THE METALLIC ORBITAL

Hypoelectronic metals are metals with a smaller number of outer electrons than available orbitals, and hyperelectronic metals are those with a larger number. The metals on the left side of the periodic table are hypoelectronic, and those on the right side are hyperelectronic. The transition metals I consider to be buffer metals; it is likely that they have to some extent the properties of either hypoelectronic or hyperelectronic metals.

Let us first consider a hypoelectronic metal, such as Aℓ. Aℓ has three valence electrons, and it has four available orbitals (3s and 3p). The aluminum atom can accordingly accept an additional bonding electron, giving it the valence 4. An aluminum atom with valence 4 forms stronger bonds with its twelve neighbors than an aluminum atom with valence 3, so that there can be expected a contraction in the lattice around the Aℓ^- atom. Accordingly there is a stabilizing interaction between Aℓ^- and the crest (maximum density) of a compressional phonon. Aℓ can accordingly be called a crest superconductor. On the other hand, with an element such as Ga (metallic valence 3.5) with more outer electrons than available orbitals, the possession of an extra electron decreases the valence by one unit. This hyperelectronic metal is accordingly a trough superconductor. Several investigators, especially Anderson[8,9] have suggested that the driving force of superconductivity is not the electron-phonon

interaction, but rather an electromagnetic interaction associated with resonance of covalent
bonds. I have pointed out[10] that with a
mechanism other than interaction with phonons
responsible for preserving the integrity of the
electric current in a superconductor, electron-
phonon interaction would be the spoiling mechanism, responsible for setting a limit on the
temperature at which the superconducting current
survives, so that the superconducting temperature
T_c would increase with decrease in the electron-
phonon interaction. Small values of the electron-
phonon interaction and high values of T_c can be
obtained by having crest and trough metals in the
same superconducting material. As expected from
this argument, alloys of two hypoelectronic metals
or of two hyperelectronic metals have values of
T_c intermediate between those for the element
metals, whereas those of a hypoelectronic metal
and a hyperelectronic metal have higher values of
T_c.

For alloys of Nb and Sn, for example, a
calculation of this sort[11] leads, by consideration only of the change in valence, to the
conclusion that the maximum value of T_c should
occur at the composition Nb_2Sn. Correction for
the strength of the bonds[12] shifts the composition
to Nb_3Sn, in agreement with experiment.

I have recently suggested[11] that metallic
conduction involves the motion of fluxons,
clusters of atoms with an extra negative charge

or positive charge, rather than of the individual atoms M^+ and M^- of the resonating-covalent-bond theory of metals. These charged clusters are expected to exist as fluctuations in the distribution of the resonating covalent bonds. In the theory they correspond to the Cooper pairs of the BCS theory of superconductivity.

A detailed theory of fluxons and resonating covalent bonds in metals has yet to be formulated, but preliminary calculations indicate that the theory is compatible with the results of experiment.

REFERENCES

1. L. Pauling, Phys. Rev., 54, 899 (1938).
2. L. Pauling, Nature, 161, 1019 (1948).
3. L. Pauling, Proc. Roy. Soc. London, A196, 343 (1949).
4. L. Pauling, J. Solid State Chem., 54, 2197 (1984).
5. B. Kamb and L. Pauling, Proc. Nat. Acad. Sci. U.S.A., 82, 8284 (1985).
6. L. Pauling and B. Kamb, Proc. Nat. Acad. Sci. U.S.A., 83, 3569 (1986).
7. L. Pauling, Proc. Nat. Acad. Sci. U.S.A., 60, 59 (1968).
8. P. W. Anderson, Mater. Res. Bull., 8, 153 (1973).
9. P. W. Anderson, Science, 235, 1196 (1987).
10. L. Pauling, Phys. Rev. Lett., 59, 225 (1987).
11. L. Pauling, submitted to Phys. Rev. Lett. (1988).
12. J. Waser and L. Pauling, J. Chem. Phys., 18, 747 (1950).

MAGNETISM AND SUPERCONDUCTIVITY IN HIGH-T_c SUPERCONDUCTORS

V. J. EMERY
Department of Physics, Brookhaven National Laboratory,
Upton, New York 19973, USA

Abstract Experimental data support a model in which the charge carriers are holes in Cu^+ sites or in O^{--} sites. T_c is proportional to the inverse square of the magnetic penetration depth (for muons), and thus to the Fermi energy for quadratic dispersion. This leads to a valid theoretical model in both the strong coupling and the weak hopping limits: a pair attraction comes from superexchange between Cu^+ holes, and from a reduction in their zero-point energy.

INTRODUCTION

Basic properties of high-temperature superconductors are deduced from microscopic experiments, paying particular attention to those of importance for the pairing mechanism.[1] The implications are discussed in terms of a model involving both copper and oxygen sites,[2] for which the charge carriers are holes in Cu^+ (filled 3d level) or O^{2-} (filled 2p level). Solution of the model in the strong-coupling limit [3,4] leads to a picture of superconductivity consistent with the experiments.

---

*   This manuscript has been authored under contract number DE-AC02-76CH00016 with the U. S. Department of Energy. Accordingly, the U. S. Government retains a non-exclusive, royalty-free license to publish or reproduce the published form of this contribution, or allow others to do so, for U. S. Government purposes.

The main features are as follows:

(1)   For sufficiently small x, both $La_{2-x}Sr_xCuO_4$ and $YBa_2Cu_3O_{6+x}$ are antiferromagnetic insulators with slightly more than one hole per unit cell in the CuO_2 planes. At low temperatures the antiferromagnetic order parameter[5] is about 50% of the classical Neel order for Cu^{2+} ions, and the form factor[6] is predominantly Cu^{2+}. The order parameter is close to the value expected for the quantum Heisenberg model,[7] which implies that the moments are localized by the strong Coulomb interactions.[1]

(2)   An exchange integral J of about 0.12 eV may be deduced[3] from the measured spin wave velocity,[8] or separately from light scattering experiments.[9]

(3)   Measurement of the magnetic penetration depth λ by muon spin resonance[10,11] provides two pieces of information:

(a)   the temperature-dependence[10] implies that there are no zeros of the energy gap.

(b)   the variation from one sample to another shows[11] that the superconducting transition temperature T_c is proportional to the low-temperature value of λ^{-2}. Using the standard relationship $\lambda^{-2} = 4\pi n_c e^2/m^{*2}$ (where e is the charge of an electron, c the velocity of light and n_c, m^* are the concentration and effective mass of the charge carriers), the measurements show that the charge is carried by the holes in excess of one per unit cell.

Property (a) is most easily explained by s-state pairing, and since n_c is proportional to k_F (where k_F is the Fermi wave vector for motion in the CuO_2 planes) whereas (b) indicates that T_c is proportional to $k_F^2/2m^*$, which is the Fermi energy for a band with quadratic dispersion. The latter implies[1-3] that pairing is produced by exchange of a high-energy excitation (greater than about 1 eV). This is the most significant known constraint on

316

he mechanism of high-temperature superconductivity: it shows that the pairing force must involve short-range spin interactions[1-3] or charge excitations,[3] or a combination of the two.

(4)   Magnetism is produced by holes mainly on the Cu sites, and superconductivity by holes on oxygen. This is demonstrated by a wealth of spectroscopic data,[12] and is supported by measurements of the magnetic form factor[6] and spin-lattice relaxation time[13] for ^{63}Cu and ^{17}O.

THEORETICAL MODEL

This last feature of the high-temperature superconductors was deduced from a model, in which the holes move on both copper and oxygen sites.[2]

The same model has been analysed in the weak hopping limit.[3,4] It has been shown[4] that the quasiparticles consist of oxygen holes, with their spin strongly correlated with that of the neighboring Cu^{2+} ions. The quasiparticles carry both spin and charge. At the bottom of the band, it is a plane-wave superposition of Cu-O-Cu spin configurations: ($\uparrow\uparrow\downarrow + \downarrow\uparrow\uparrow - 2\uparrow\downarrow\uparrow$), which may also be obtained by diagonalizing the strong Cu-O superexchange Hamiltonian.

When two quasiparticles in a singlet state come close together, there is an attractive interaction[3] due: (a) to superexchange between the constituent Cu-holes, enhanced by the Coulomb interaction between Cu-holes and O-holes, (b) to reduction of the zero-point energy of Cu-holes. The former is a magnetic interaction, the latter involves only charge fluctuations. Both have intermediate states with high (electronic) energies and lead to T_c proportional to the Fermi energy. These interactions are effective for 6 first neighbors and 16 second neighbors of an oxygen hole. The large number of neighbors, and the enhancement of superexchange are essential for overcoming the Coulomb interaction, and obtaining a net attraction. The scattering amplitude for relative angular momentum is proportional to n for small n_c. Thus s-state pairing (-O) will dominate for the low carrier concentrations of the high-T_c superconductors.

ACKNOWLEDGEMENT

This work was supported by the Division of Materials Science, U. S. Department of Energy, under contract DE-AC02-76CH00016.

REFERENCES

1. This analysis was first given in V. J. Emery, Nature 328, 756 (1987), but many more corroborating experiments have been carried out since that time.
2. V. J. Emery, Phys. Rev. Lett. 58, 2794 (1987).
3. V. J. Emery and G. Reiter, Phys. Rev. B.
4. V. J. Emery and G. Reiter; to be published.
5. Y. J. Uemura et al., Phys. Rev. Lett. 59, 1045 (1987).
6. T. Freltoft et al., Phys. Rev. B 37, 137 (1988); C. Stassis et al., to be published.
7. J. Oitmaa and D. D. Betts, Can. J. Phys. 56, 897 (1978); J. D. Reger and A. P. Young, Phys. Rev. B 37, 5978 (1988).
8. G. Shirane et al., Phys. Rev. Lett. 59, 1613 (1987); G. Aeppli (Private Communication); M. Sato et al., (Preprint).
9. K. B. Lyons et al., Phys. Rev. Lett. 60, 732 (1988).
10. D. R. Harshman et al., Phys. Rev. B 36, 2386 (1987).
11. Y. J. Uemura et al., Phys. Rev. B, July (1988).
12. J. M. Tranquada, S. M. Heald, A. R. Moodenbaugh, and M. Suenaga, Phys. Rev. B 35, 7187 (1987); J. M. Tranquada, S. M. Heald, and A. R. Moodenbaugh, Phys. Rev. B 36, 5263 (1987); See also papers in Novel Superconductivity, S. A. Wolf and V. Z. Kresin (Plenum, N. Y., 1987); and High Temperature Superconductors, S. Lundqvist, E. Tosatti, M. P. Tosi , and Yu Lu (World Scientific: Singapore, 1987).
13. Y. Kitaoka, K. Ishida, K. Asayama, H. Katayama-Yoshida, Y. Okabe, and T. Takahaski (Preprint).

SPECULATIONS ABOUT INHOMOGENEITY IN THE HIGH-T_C SUPERCONDUCTORS

MORREL H. COHEN
Exxon Research and Engineering Company, Annandale, New Jersey 08801

Abstract In the temperature dependence of the physical properties of the high temperature superconductors, there is evidence for the existence of three characteristic temperatures $T_O > T_S > T_C$, where T_C is the conventional superconducting transition temperature. The evidence is briefly discussed, and speculations are put forward concerning two possible origins from inhomogeneity of the additional temperatures T_O and T_S. One proposed origin is in granularity derived from the interplay between disorder and anisotropy in the superconductivity, and the second is in an inhomogeneous metal-nonmetal transition. We conclude that the facts favor the latter explanation.

I. INTRODUCTION

When one examines the experimental data on the high temperature superconductors, one sees over and over again three characteristic temperatures, instead of a single superconducting phase transition temperature. One sees these three temperatures in the temperature dependence of the transport properties, e.g., the

resistivity[1] and thermopower;[2] of the ultra-
sonic properties,[3] e.g., the velocity and
damping; of the thermodynamic properties, e.g.,
the specific heat;[4] of the lattice parameters;[5]
and of the magnetic properties, e.g., the
susceptibility[6] and the EPR.[7]

Typical temperature dependences for the
resistivity[1] and thermopower[2] are sketched in
Figures 1 and 2, respectively, for materials in
the $(La,Sr)_2CuO_{4-y}$ system (214) and the
$YBa_2Cu_3O_{7-y}$ system (123). The resistivity and
thermopower both show linear temperature
dependences above T_O, where a transition
towards less metallic behavior occurs. Both
properties vanish over a narrow transition
region, centered about a nominal superconducting
temperature T_C. However, the earliest evidence
of departure from the behavior below T_O occurs
at a clearly marked temperature T_S, well above
T_C.

None of the three temperatures marks a
sharp, macroscopic phase transition. T_C and T_S
mark transitions of comparable widths, but that
at T_O is considerably broader.

Most of the data we have reviewed for
evidence of T_S and T_O, as well as T_C, were
taken on ceramic samples. Some, however, comes
from single crystal samples.[8]

Values of T_C, T_S, and T_O are given in
Table I.

TABLE 1

Material	T_C	T_S	T_O
214	30-40K	100K	200K
123(7)	80-95K	110-130K	200-300K

In addition, T_S tracks T_C in its dependence on y in the 123(7-y) system.[6]

The properties are quite sensitive to oxygen deficiency. The resistance peak in the 214 material shown in Figure 1 is systemati-

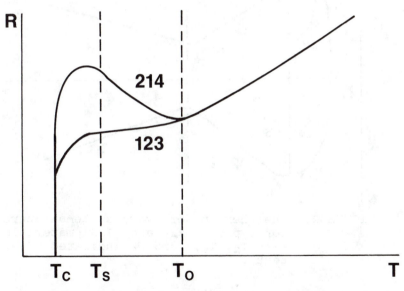

FIGURE 1. Electrical resistance versus temperature for typical 214 and 123 materials with T_C, T_S, and T_O indicated (see text).

cally reduced, as the oxygen deficiency is eliminated, until the temperature dependence resembles that of the 123 material.[9] Similarly, in the 123 material, the break in slope shown in Figure 2 at T_O becomes less marked.[9] Nevertheless, while the values of T_O change with oxygen deficiency, the change in behavior at T_O, while weakened, persists. T_S, on the other hand, shows little change.

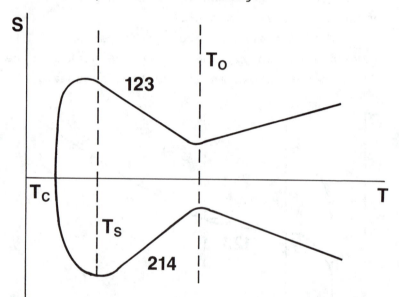

FIGURE 2.   The thermopower versus temperature for typical 214 and 123 materials with T_C, T_S, and T_O indicated (see text).

These anomalies raise a puzzle. What is happening at T_S and T_O? We address this question in the next two sections of this paper, speculating about the consequences of inhomogeneity for the physical properties. The inhomogeneities we consider are fluctuations and granularity in anisotropic, disordered superconductors in Section II, and inhomogeneous metal-insulator transitions in Section III. We conclude, in Section IV, that the physical picture provided by an inhomogeneous metal-insulator transition provides the more probable explanation of the origin of T_O and T_S.

II. THE DEVELOPMENT OF GRANULARITY IN DISORDERED ANISOTROPIC SUPERCONDUCTORS

In the present section, we explore the possibility that superconducting gap fluctuations start at T_O. We show that the existence of an intermediate temperature T_S, above the bulk transition temperature T_C, then follows in a superconductor which is both sufficiently disordered and sufficiently anisotropic. At T_S, the material becomes a granular superconductor, containing disjoint superconducting regions with mutually incoherent phases. At T_C, the phases lock, and the bulk superconducting transition occurs. We begin with a brief discussion of gap fluctuations.

A. GAP FLUCTUATIONS

The density of states $N(E)$ in a normal metal varies smoothly with energy E through the Fermi

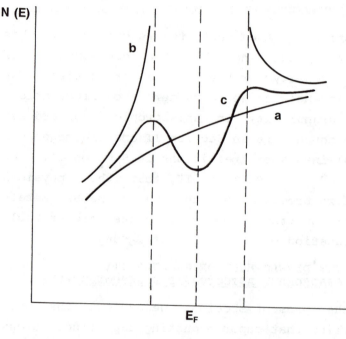

FIGURE 3.   Density of states N(E) as a function
of single quasiparticle energy E in
the vicinity of the Fermi energy E:
a. normal metal; b. BCS superconductor below T_C; c. in the presence of
superconducting   gap   fluctuations
above T_C.

energy E_F, as in Figure 3, curve a. Below T_C,
in a BCS superconductor, a gap 2Δ opens at E_F
in the density of states, as in Figure 3, curve
b. However, there are circumstances in which
the density of states does not change abruptly
from that of the normal metal to that of the

superconductor, developing a gap bounded by singularities in $N(E)$ at T_C. This can occur when the mean-field approximation inherent in the BCS theory and its various generalizations breaks down, as in sufficiently anisotropic materials, or materials of reduced dimensionality, and gap fluctuations occur over a substantial temperature range above T_C.

If such gap fluctuations were to set in at T_O, and grow as T decreases further, the gap parameter Δ would become position and time-dependent,

$$\Delta = |\Delta(x,t)| e^{i\phi(x,t)} \tag{2.1}$$

Below T_O, the magnitude of Δ would have a finite time average value

$$\bar{\Delta} = \overline{|\Delta(x,t)|}^t \neq 0 , \tag{2.2}$$

but Δ itself would vanish because of the phase fluctuations,

$$\overline{e^{i\phi(x,t)}}^t = 0 . \tag{2.3}$$

There would be no macroscopic superconductivity, or even superconducting fluctuation phenomena, if the fluctuations were sufficiently

325

rapid, but the density of states would show a pronounced minimum at E_F, as in Figure 3, curve c.

As a consequence of the increasing dip in the density of states, i.e., the gradual development of a gap, the resistivity increases, the magnitude of the thermopower increases, and the magnetic susceptibility decreases, as, in fact, is observed below T_O. It is consistent to infer from the data that T_O marks the onset of gap fluctuations. It is important to remember that such a reduction in the density of states at E_F would arise from any kind of gap fluctuation, not only from superconducting gap fluctuations. Any other source of such fluctuations, however, would be associated with a metal-nonmetal transition, as we explore in Section III.

B. GRANULAR SUPERCONDUCTIVITY

A typical granular superconductor consists of metallic grains separated by barriers of some sort. Above a bulk transition temperature, $T>T_B$, the grains are normal. Below T_B, $T_B>T>T_C$, where T_C is the bulk transition temperature, the grains are individually superconducting. They are separated by weak links, if the barriers are normal metallic, or by Josephson junctions, if the barriers are insulating. There is phase coherence within the grains, but the phase is random between grains, so that there is no bulk superconductivity. As a consequence, certain characteristic behaviors occur: (1) conductivity anomalies of the form

$$\Delta\sigma \propto -\ln \left| (T - T_c)/T_c \right| , \qquad (2.4)$$

(2) anomalous magnetotransport effects,[10] and (3) tunnelling or weak link phenomena.[10]

At T_c the phases lock, and there is bulk superconductivity below T_c with, however, additional behaviors not found in homogeneous bulk superconductors: (4) incomplete Meissner effect,[10] (5) glassy relaxation,[10] (6) low critical current J_c,[10] and (7) internal Josephson phenomena.[10]

The high-temperature superconductors are found to behave like granular superconductors. In the temperature range $T_c<T<T_s$, phenomena 1-3 above are observed. In the range $T<T_c$, effects 4-7 above are observed. The temperature dependence of the excess conductivity of a 123 sample relative to a base line obtained by extrapolation from above T_s is shown in Figure 4. It is accurately fitted by Eq.(2.4) for $T_c<T\leq T_s$.[1]

One can also fit the excess conductivity to the conventional theory of fluctuations in a homogeneous anisotropic superconductor. However, there is then no evidence of the 2D-3D crossover predicted by the theory.[11]

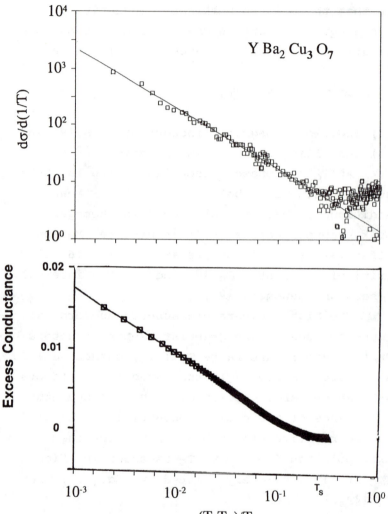

FIGURE 4. Excess conductivity $\Delta\sigma$ of a sample of $YBa_2Cu_3O_7$ below $T_S = 125 \pm 5$ (courtesy J. Stokes): the baseline from which $\Delta\sigma$ is determined by subtraction is obtained by extrapolation of a linear dependence from above T_S. a. $\Delta\sigma$ vs. $\log[(T-T_c)/T_c]$. b. $\log d\sigma/d(1/T)$ vs. $\log[(T-T_c)/T_c]$. The slope of the straight line is -1.

328

To summarize the proposals of Section IIA and the present section, the material is a normal metal of unusual properties above T_O. Gap fluctuations occur between T_O and T_S, but phases remain uncorrelated. Granular super-conducting grains with uncoupled phases emerge between T_S and T_C. The material becomes a bulk superconductor below T_C with persistence of granularity.

C. ANISOTROPY

All high temperature superconductors contain CuO_2 planes.[12] In each plane, the mean Cu valence is between 2 and 3, or the oxygen charge is smaller than 2, or both.[13] It is now well established that these planes are the primary source of the superconductivity.[14] The planes are stacked singly, separated by other constituents which build up the structure and control the conduction electron content of the planes, or stacked in relatively closely coupled pairs, triplets, etc.

The charge carriers now appear to be holes primarily on the oxygen ions. The holes move easily and interact strongly within the CuO_2 planes, but move with difficulty, and interact weakly between planes. These superconductors are thus highly anisotropic in their electronic properties as well as their atomic structures.

D.   TWO-DIMENSIONAL SUPERCONDUCTORS

Suppose, to begin with, we ignore the coupling between planes. We would then be dealing with two-dimensional superconductors. The super-conducting electron density is sketched for a typical 2D superconductor[15] in Figure 5. N_S is

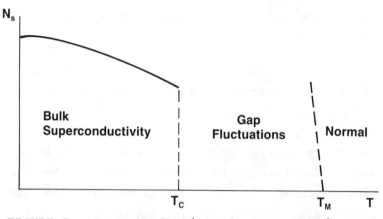

FIGURE 5.   Superconducting electron density N_S vs. temperature T. T_M is the mean-field transition temperature, and T_C is the bulk transition temperature.

zero above T_M, where the material is a normal metal. Between T_C and T_M there are gap fluctuations, as discussed in Section IIA, but unbound 2D supercurrent vortices kill the macroscopic phase coherence. At T_C, the Kosterlitz-Thouless transition temperature, N_S jumps from zero to a universal value, as the supercurrent vortices bind in pairs of opposite sense. The material is a bulk superconductor below T$_c$ even though the phase coherence does not persist to infinity, but falls off algebraically with distance.

E. EFFECT OF 3D COUPLING

The 2D case shows a mean-field transition temperature T$_M$, which could be assimilated to T$_c$ and a T$_C$, but T$_S$ is absent. Suppose, therefore, we put back the coupling between planes, measured by an appropriate coupling constant, K$_\perp$. The effect on T$_C$ and T$_M$ is shown in the phase diagram[16] of Figure 6. For vanishing K$_\perp$, the temperature axis is a line of critical points extending from OK to T$_C$<T$_M$. As K$_\perp$ increases, the T$_C$ trajectory emerges from the T axis with zero slope, ultimately becoming linear in K$_\perp$ when the 2D-3D crossover is complete. Similarly, the T$_M$ trajectory emerges linearly from the T axis, and merges with the T$_C$ trajectory, when crossover is complete, if the coherence length is large enough, as shown in Figure 6. However, the coherence length for the high temperature superconductors (HTSC) may not be large enough for such a merger when T$_C$ exceeds about 30K, and T$_M$ may run parallel to T$_C$ with a fluctuation region between, T$_C$<T<T$_M$.

In either case, there is a fluctuation region bounded by two characteristic temperatures, T$_C$ and T$_M$, for some range of K$_\perp$, but there is no evidence of a T$_S$. What is missing is the

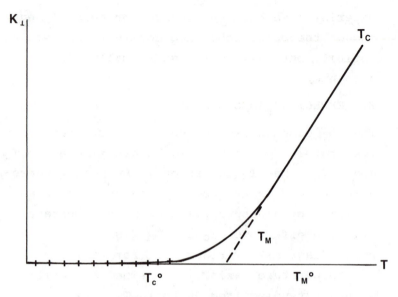

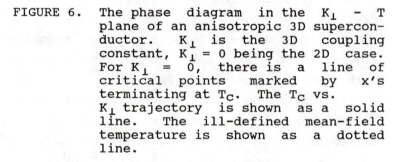

FIGURE 6. The phase diagram in the $K_\perp$ – T plane of an anisotropic 3D superconductor. $K_\perp$ is the 3D coupling constant, $K_\perp = 0$ being the 2D case. For $K_\perp = 0$, there is a line of critical points marked by x's terminating at T_C. The T_C vs. $K_\perp$ trajectory is shown as a solid line. The ill-defined mean-field temperature is shown as a dotted line.

disorder present in almost all materials prepared to date, the consequences of which we examine in the next section.

F. EFFECTS OF DISORDER

The materials prepared to date contain, in varying degrees, structural disorder manifested as (1) grain boundaries or extended defects or

(2) compositional disorder associated with random atomic substitution, oxygen vacancies, spinodal decomposition, etc. Suppose that the disorder introduces random local variations in the mean-field transition temperature T_M. Over a range of temperatures around some temperature T_O, both of which depend on the probability distribution of T_M, localized regions then emerge in which gap fluctuations occur.

In the temperature range $T_S < T < T_O$, these regions increase in number, and grow individually in size. Within them, Δ is nonzero, but the phase ϕ is random. They are anisotropic in shape, with much greater extent, $l_{||}$, within the planes, than in the stacking direction, $l_\perp$. Such localized regions of gap fluctuation are potentially the grains of a granular superconductor.

Suppose, now, that T_S is the superconducting phase transition temperature within these grains. That is, we are supposing that T_O is associated with the T_M, and T_S with the T_C of Figure 6, and that, because of both the anisotropy (quasi-2D character) and the small coherence length, there is a substantial separation between the T_M and T_S of Figure 6. Thus, the localized regions of finite Δ phase lock internally, and become superconducting grains. These grains couple through the weak links provided by the not yet superconducting matrix, but above T_C the grain phases remain incoherent. T_C is either the bulk transition temperature at which the grain phases lock, or the temperature at which the locally-super-conducting volume fraction reaches the percola-tion threshold, whichever occurs at higher T. In either case, non-superconducting regions and isolated grains remain below T_C, gradually disappearing as T is lowered.

G. DISCUSSION

The picture developed in the previous section contains the three characteristic temperatures. T_O lies in the upper range of the local mean-field transition temperatures T_M. T_S is the internal bulk transition temperature within typical large grains, which form in the range $T_S<T<T_O$. T_C is either a phase-locking transi-tion temperature, or a true bulk transition temperature, depending on whether the volume fraction of material with $T_M \geq T_S$ reaches the percolation threshold before phase locking occurs.

This picture has interesting consequences. Consider first the repeated observation of large gaps Δ such that $2\Delta/kT_C > 3.5$, the BCS weak-coupling value. As there is other evi-dence, which can be interpreted as supporting weak coupling,[17] this poses a puzzle. In the present picture, the large gaps would be associated with T_O, i.e., $2\Delta = 3.5\ kT_O$ not $3.5\ kT_C$. This is quite interesting, in connection with recent gap determinations via NQR spectroscopy.[18] Two values of $2\Delta/3.5k$ have been found in 123 specimens, 200K, associated with the planes, and 60K, associated with the chains. We note that 200K is a typical value of T_0 for the planes, and suggest that the 60K gap is parasitic on the gap in the planes.

Suppose one were to eliminate the disorder in these materials. If the above picture were correct, their T_C would then move up to T_S, e.g., increase from 90K to 125K in the 123 materials. The phase diagram of Figure 6 would then apply. Next, by strengthening the coupling between the planes, T_S would move up towards T_0, as T_0 itself would increase. For example, $T_C = 84K$ for the Tl containing materials, in which only single CuO_2 planes occur, but 106K for the materials containing pairs of planes, and 124K for triple planes.[19] Also, T_0 can be increased within a given structure by varying the composition. For example, in $YBa_2Cu_3O_{7-y}$, T_0 increases from 200K to above 300K as y decreases. Since a T_0 of 320K has been observed, the present picture implies that a room-temperature superconductor is not an idle dream.

Unfortunately, there are serious shortcomings in the present picture. While there is a strong case for inhomogeneity and granularity in the superconducting properties, there is no real evidence that the gap fluctuations indicated by the data between T_S and T_O are associated with a superconducting gap. Moreover, why should T_O vary from 210K to 320K upon oxygenation of 123(7-y) (0.05>y>0), while T_S and T_C do not change?[1] In either interpretation of T_C, one would expect it to be a very sensitive parameter in the theory. Similarly, why, in the 214 materials, should oxygenation change the temperature dependence of the electrical resistivity, from activated between T_S and T_O to 123-like, without any accompanying changes in T_S and T_C?[9] Perhaps the most serious problem is the relative insensitivity of T_C to sample preparation, e.g., similar values of T_C in single crystal and ceramic specimens of similar composition.

We therefore ask whether there is an alternative, perhaps intrinsic, source of inhomogeneity other than disorder. We answer this question in the affirmative in the next section.

III. THE POSSIBILITY OF AN INHOMOGENEOUS METAL-NONMETAL TRANSITION

The T-x phase diagram of the prototypical 214 material, $La_{2-x}Sr_xCuO_{4-y}$, is sketched in Figure 7 for x small, but nonzero. For small x, there is an antiferromagnetic phase, with a Neel temperature T_N, which decreases rapidly with increasing x.[20] Above T_N, there is a persistence of 2D antiferromagnetic order, with a

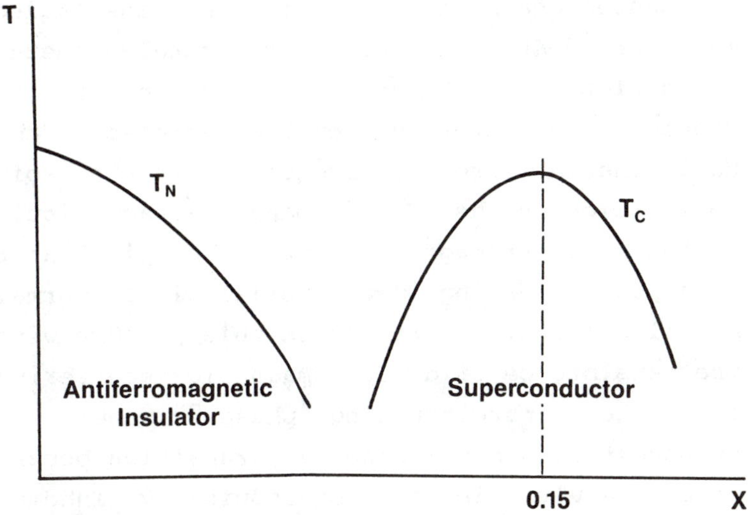

FIGURE 7. Temperature-composition phase diagram for La$_{2-x}$Sr$_x$CuO$_4$ (sketch).

long coherence length ξ.[21] There is a gap on the x-axis between the antiferromagnetic phase and the superconducting phase. The superconducting transition temperature, T$_C$, peaks at an x of 0.15. Within the superconducting phase, the diffuse magnetic scattering characteristic of the 2-D antiferromagnet persists, with ξ decreasing, and the susceptibility χ increasing with x.

The Hall constant R$_H$ shows striking evidence of a metal-nonmetal transition at the same value of x at which superconductivity occurs.[22] At lower values of x, R$_H \propto 1/x$, as would be expected if Sr doping of La$_2$CuO$_4$

introduces one hole per Sr into an insulating material. At the apparent metal-nonmetal transition, the abrupt increase in carrier concentration, inferred from the decrease of the Hall constant, strongly suggests that the spin correlation in the $d_{x^2-y^2}$ band characteristic of the antiferromagnetic Mott-Hubbard insulator disappears, leaving the material in a normal metallic state. To reconcile this picture with the persistence of diffuse magnetic scattering into the superconducting phase,[23] one can suppose that a metal-nonmetal transition occurs, which results in a microscopically inhomogeneous material, part normal metallic, which becomes superconducting, and part Mott-Hubbard insulator, which remains a 2D antiferromagnet with finite coherence length.

Figure 8 shows a sketch of the T-y phase diagram of the prototypical 123 material $YBa_2Cu_3O_{7-y}$. It has features similar to those of Figure 7. For small y, the material is superconducting. For larger y, it is an antiferromagnetic insulator. There is a gap in y, between the superconducting and the antiferromagnetic phases, with some evidence of a first-order phase transition there.

In both materials the superconductivity is entirely (214), or primarily (123) in the planes. The nominal valence z of the in-plane Cu ions is 2.15 at y=0.[19] Thus, T_C is maximal

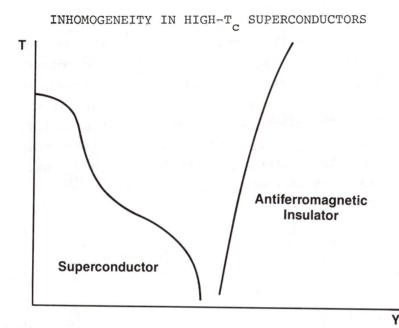

FIGURE 8. Temperature-composition phase diagram for $YBa_2Cu_3O_{7-y}$ (sketch).

in both materials, when the valence z of the
in-plane Cu atoms is 2.15.

Because of these strong similarities, we
shall not discuss the materials separately, but
simply examine the properties of the CuO_2
planes as functions of z. Suppose we ignore,
for the moment, the fact that variations of z
imply deviations from charge neutrality in a
material of fixed composition. Let us specu-
late that, below some temperature T^*, there are
at least two locally stable values of z for
CuO_2 planes, z=2 and z=2.15, corresponding
respectively to x=0 in the 214 material, and

y=6.5 in the 123 material, on the one hand, and
to x=0.15 in 214, and y=0 in 123 on the other.
The corresponding free energy is sketched in
Figure 9 for T>T* and T<T*. Continuing to
ignore the question of electrical neutrality,
such a free energy would imply a first-order
phase transition, a metal-insulator phase
transition corresponding to the phase diagram
shown in Figure 9.

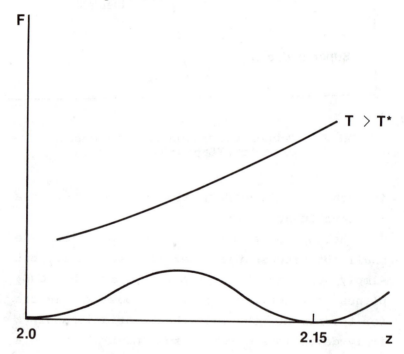

FIGURE 9. Hypothetical dependence of free
energy on nominal Cu valence for
CuO_2 planes, assuming uniformity, and
ignoring deviations from electrical
neutrality.

However, despite the existence of a first-order phase transition, a macroscopic phase separation would be prevented by deviations from local electrical neutrality. Consider $La_{2-x}Sr_xCuO_4$, the simpler material. If one adds the electrostatic potential energy to the free energy F sketched in Figure 9, supposing nothing more than F itself being symmetric in z about z=2.075, and that the distance scale of the resulting inhomogeneities is large, compared to the healing length, so that one can introduce interfacial free energies, one finds that a system, in the two-phase region of Figure 8, does indeed undergo an inhomogeneous metal-nonmetal transition, with a change in morphology occurring at x=0.075, as shown in Figure 10.

For x<0.075, the material consists of disjoint insulating regions, separated by a continuous metallic matrix, as shown in Figure 11. One supposes that the superconductivity characteristic of the bulk metallic material with z=2.15 or x=0.15 is killed by the proximity effect and phase fluctuations. For x>0.075 on the other hand, the material consists of disjoint metallic regions separated by insulating regions. The insulating regions would have z=2, and would therefore be 2-D antiferromagnetic, with a coherence length decreased by disorder and size effects, consistent with the observations.[23]

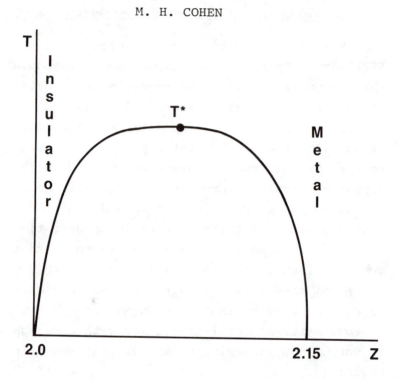

FIGURE 10. The phase diagram following from the free energy of Figure 9.

The three characteristic temperatures emerge quite naturally for the latter case. T_O is the microscopic phase separation temperature for that value of x, or more generally x-2y. T_S is the bulk transition temperature of the metallic regions. T_C is the phase-locking temperature. In the range $T_S < T < T_O$, the insulating barriers between the metallic domains provide the energy gap or gap fluctuations required by the transport properties. In the range $T_C < T < T_S$ and below T_C, one has a granular

X < 0.075   X > 0.075

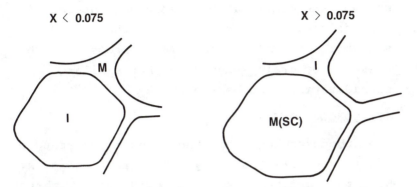

FIGURE 11. The morphology of the mixed - phase
            region.   For   x<0.075,   disjoint
            insulating regions (I) are embedded
            in a metallic matrix (M).   For
            x>0.075, the metallic regions are
            disjoint and embedded in an insulat-
            ing   matrix.   Superconductivity
            occurs for x>0.075.

superconductor.   The   material   is,   ideally,
homogeneous for   x=0.15.   In   general,   for
x−2y>0.075,   one   is   in   the   superconducting
region of the phase diagram. For x ≃ 0.15, as
y is decreased,   the inhomogeneities   decrease.
In particular,   the widths of the barriers
decrease, reducing the deviation of the   transport properties in the   range T$_S$<T<T$_O$ from   the
extrapolations of their values from above T$_O$,
as is observed.

    This picture has some very attractive
features for the 214 systems, but it is too
early to evaluate its relevance.   The   123

343

systems are complicated by spinodal decomposition of the oxygen concentration, when $y \neq 0$, as are both the 123 and 214 systems by the presence of grain boundaries, but similar considerations can be applied.

IV. CONCLUSIONS

The phenomenology of the high temperature superconductors is very rich and complex. Within it one can see clear evidence of three characteristic temperatures. Any theories of high-temperature superconductivity must be capable of explaining their presence. To do so, the theories must deal explicitly with the anisotropy, the disorder, and the proximity of metal-insulator transitions in these materials.

In the present set of ideas, there are two possible interpretations of T_O. It can be associated with a randomly varying mean-field transition temperature (Section II), or a crossing of the coexistence curve of a microscopically inhomogeneous metal-insulator transition (Section III). In both cases, T_S is associated with superconductivity within local regions, which play the role of grains in a granular superconductor, and T_C is the phase-locking temperature.

It would be premature to say that the preponderance of evidence favored one or the other interpretation. It is fair to say, however, that the metal-insulator transition picture is freer of objections and consistent with a larger range of facts. If that picture proves correct, many of the microscopic theories proposed thus far would be irrelevant. In particular, there would be no need to suppose that the electronic structure of the superconductor regions involved strong correlations generating local spins, and to base the theory of superconductivity on their existence. Prior experience with materials of reduced dimensionality, indeed, mitigates against the persistence of such strong correlations, as screening is increased by doping.[24]

V. ACKNOWLEDGEMENT

I am deeply grateful to my colleagues Drs. S. Bhattacharya, J. Stokes, A. N. Bloch, G. S. Grest, S. K. Sinha, T. Witten, A. J. Jacobson, J. Newsam, and D. Johnston for their encouragement and skepticism, and for their ideas and data.

REFERENCES

1. A. J. Jacobson, et al., in "Chemistry of Oxide Superconductors", edited by C. N. R. Rao (Blackwell, Oxford) (In Press).
2. M. F. Hundley, et al., Phys. Rev. B35, 8800 (1987); R. C. Yu, et al., Phys. Rev. B37, 7963 (1988).
3. S. Bhattacharya, et al., Phys. Rev. B37, 5901 (1988); S. Bhattacharya, et al., Phys. Rev. Lett. 60, 1181 (1988).
4. T. Laegrid, et al., Nature 330, 637 (1987).
5. P. M. Horn, et al., Phys. Rev. Lett. 59, 2772 (1987); Linhai Sun, et al., Phys. Rev. B (September 1, 1988) (In Press).
6. D. C. Johnston, et al., Proc. Int. Conf. on High Temperature Superconductors and Materials and Mechanisms of Superconductivity, Interlaken, Switzerland, Physica B (1988) (In Press).
7. H. Thomann, et al., Phys. Rev. B (In Press).
8. N. P. Ong, et al. (Preprint).
9. J. P. Stokes (Unpublished).
10. B. Abeles, Appl. Sol. St. Sci. 6, 1 (1976).
11. N. P. Ong - this volume.
12. D. W. Murphy, et al., Phys. Rev. Lett. 58, 1888 (1987); C. C. Torardi, et al., submitted to Science; D. R. Vehlen, et al., Nature, 332, 334 (1988).
13. A. J. Jacobson and J. M. Newsam, unpublished bond-length analysis; G. Xiao, et al., Phys. Rev. Lett. 60, 1446 (1988).
14. There are many indications. One of the more elegant is the demonstration of the absence of anisotropy of the critical current parallel to the planes, M. Oda, et al., Phys. Rev. B38, 252 (1988).
15. B. I. Halperin, Proc. Kyoto Summer Inst. on Low Dimensional Systems, September 8-12, 1979; B. I. Halperin and D. R. Nelson, J. Low Temp. Phys. 36, 599 (1979).

16. cf. J. V. Jose, et al., Phys. Rev. B16, 1217 (1988).
17. R. W. Cohen, M. H. Cohen, and M. L. Cohen, submitted; M. B. Salomon and J. Bardeen, Phys. Rev. Lett. 59, 2615 (1987).
18. W. W. Warren, et al., Phys. Rev. Lett. 59, 1860 (1987).
19. S. S. P. Parkin, et al., Phys. Rev. Lett. 60, 2539 (1988).
20. S. K. Sinha, MRS Bulletin, Vol. 13, No. 6, 24 (1988).
21. G. Shirane, et al., Phys. Rev. Lett. 59, 1613 (1987).
22. N. P. Ong, et al., Phys. Rev. B35, 8807 (1987).
23. R. J. Birgeneau, et al. (Preprint).
24. A. N. Bloch and S. Mazumdar, J. de Physique 44, C1273 (1983).

ENERGY GAPS IN A LOCAL-PAIR SUPERCONDUCTOR

I.O.KULIK
Institute for Low Temperature Physics
and Engineering, Ukr. SSR Acad. Sci.
47 Lenin Ave.,310164 Kharkov,U.S.S.R.

Abstract A Local-pair model of high-T_C
superconductivity is considered. The model
predicts two gaps, the tunneling one which
is strongly anisotropic,$(\Delta = \Delta_k)$, and large
compared to BCS value, and the pair-gap, R,
which has an anomalous temperature dependence with the ratio $2R(0)/T_c \sim 2$.

Among the theoretical models describing high-
T_c superconductivity, those with ready (existing
above T_c) electronic pairs are of specific interest. Anderson's theory for high-T_c[1,2] can be considered as belonging to this type, since the RVB state
formation with bose-like spinon/holon pairs pre-
cedes superconductivity. We have proposed a model[3]
based on oxygen-bound pairs (figuratively speak-

ing, electronic molecules localized   at   oxygen
atom(s)) interacting with each other indirectly
through the exchange of conduction electrons.

This local-pair model of high-T_c[3] predicts specific (and experimentally verifiable features of a
superconductor in the normal state. The chemical
potential μ is shown to have a linear   temperature
dependence (followed by a constant value for $T<T_c$)

349

$$\mu = E_o - \frac{kT}{2} \ln \frac{1-\gamma}{\gamma} \tag{1}$$

which shows up as a large (of the order of $10^2 \mu V/K$) contribution to the contact potential of high-T_c material. The other property is a weak temperature dependence of the thermoelectric power

$$S = \frac{k}{2e} \frac{\sigma_p}{\sigma_s + \sigma_p} \ln \frac{1-\gamma}{\gamma} \tag{2}$$

and a linear temperature dependence of the resistivity $\rho \sim mT/ne^2\hbar$. Here σ_p (σ_s) is the pair (single electron) conductivity, γ the pair occupancy per local site. Note that the prefactor in Eq.(2) equals $43 \mu V/K$. Under the condition $\sigma_p > \sigma_s$ formula (2) (as well as (1)) permits the direct determination of the pair density in a metal.

In a specific high-T_c material, $YBa_2Cu_3O_{7-x}$, local pairs were supposed [4] to be associated with CuO chains, and the conduction electrons with CuO_2 layers. Purely repulsive interactions resulted in local pair formation, due to the disproportionation reaction

$$O^{2-}Cu^{2+}O^{2-} + O^-Cu^{2+}O^- \leftrightarrow O^{2-}Cu^{2+}O^- + O^-Cu^{2+}O^{2-}$$

where O's are O(4), according to the conventional atom labelling scheme [5].

An NQR experiment[6] revealed different characte-ristic energy gaps, corresponding to chains and layers. These can be described within the two-component local-pair scheme [7]. The BCS order parameter for conduction electrons, Δ_K, and the

350

order parameter for local pairs, $\square = \langle A_j \rangle$ (A_j is the annihilation operator for the pair) obey the self-consistency equations

$$\Delta_k = V_{kk}\frac{\square}{2R}\tanh\beta R \ , \quad \gamma = \frac{1}{2}(1-\frac{E}{R}\tanh\beta R\,)$$

$$\tag{3}$$

$$\square = \frac{1}{N}\sum_k V_{kk}\frac{\Delta_k}{2\mathcal{E}_k}\tanh\frac{\beta\mathcal{E}_k}{2}, \quad \beta = 1/T, \quad R = \sqrt{(E_o-\mu)^2+\square^2}$$

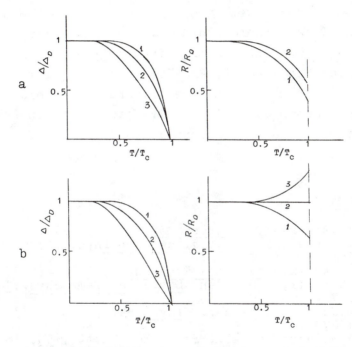

FIGURE 1. Temperature dependence of the gap in single-electron spectrum (Δ) and in the pair spectrum (R). (a)γ=0.25. Curves 1,2 ,3 correspond to the values of the electron-pair coupling constant λ = 0.54; 0.12; 0.03; (b) γ =0.1. Curves 1,2,3 correspond to λ =0.3; 0.05; 0.01.

where ε_k is the BCS energy $\sqrt{\xi_k^2 + \Delta_k^2}$, and V_{kk} is the pair-electron interaction potential (hybridization energy).

As follows from Eqs.(3), the relation between Δ_k and V_{kk} is local, rather than integral. Therefore, one can expect a large anisotropy of the gap, which probably accounts for the wide spread of gap values obtained in tunneling experiments. The quantity R can be considered as a pair-gap since, according to Ref. 7, the absorption spectrum of a high-T_c superconductor has a logarithmic singularity at $\hbar\omega = 2R$. Other singularities arise at energies $\hbar\omega = 2\Delta - 2R$, 2Δ (BCS-type singularity), and $2\Delta + 2R$. Δ and R are temperature-dependent in the manner shown in Fig.1. Note that in the isotropic case $(V_{kk} = \text{const})$ $2R_0/T_c$ is of the order of 2.

REFERENCES

1. P.W.Anderson, Science, 235, 1196 (1987).
2. I.I.Ukrainskii, Sov.J.Low Temp.Phys., 13, 883 (1987).
3. I.O.Kulik, Sov.J.Low Temp.Phys.,13,879 (1987).
4. I.O.Kulik, A.G.Pedan, Sov.J.Low Temp.Phys.,14, No.7 (1988).
5. J.D.Jorgensen et al., Phys. Rev., B36, 3608 (1987).
6. W.W.Warren et al., Phys. Rev. Lett., 59, 1860 (1987).
7. I.O.Kulik, Sov.J.Low Temp.Phys.,14,209 (1988).

A FUNCTIONAL INTEGRAL METHOD FOR THE STRONGLY CORRELATED HUBBARD MODEL

SANJOY K. SARKER
Department of Physics & Astronomy, The University of
Alabama, Tuscaloosa, Alabama 35487-1921

Abstract   A functional integral method is presented, in
which the hopping term in the strongly correlated
Hubbard model is generated via a coupling to an
auxillary Fermi field. Upon integration of the original fields, the interactions among the new fermions
correspond to the number of hops, and are characterized
by momentum and frequency-dependent coupling constants.
The lowest-order term gives rise to the expected two-
band structure of the Green's function. For large U
and N=1, higher-order terms can lead to super-exchange
in the original system, and can thus produce pairing
through the resonating valence bond scheme of Anderson.

It has been suggested that pairing in high-T_c superconductors[1] arises from a purely electronic mechanism.[2,3]
Anderson[2] has proposed a resonating-valence-bond (RVB) pairing in high-T_c superconductors based on the strongly
correlated Hubbard model which is described by the
Hamiltonian

$$H = t\Sigma\varepsilon(k)c_{k\sigma}^{+}c_{k\sigma} + U\Sigma_{r} n_{r\uparrow}n_{r\downarrow} \tag{1}$$

Here c_σ destroys a particle of spin σ at the lattice site r,
U is the on-site repulsive energy. The hopping from site to
site gives rise to a band with energies $t\varepsilon_k$ and bandwidth
proportional to t.

   In the strongly correlated regime, U is very large, and

when the density n is close to unity, hopping is controlled
by the local environment. Then, apart from direct hopping,
two or more particle hopping-exchange processes become important, giving rise to strong spin-fluctuations. In this
limit, the conventional perturbation scheme breaks down.

Here we present a functional integral method in which
the hopping-exchange processes appear naturally. We write
the partition function Z as a functional integral[4] over the
anti-commuting Grassmann variables $c(r\tau)$ and $c^*(r\tau)$ where
$0 < \tau < \beta = (kT)^{-1}$. The main purpose of the present scheme
is to treat the local U term in (1) exactly, and the hopping
term perturbatively. The key element in our method is the
identity

$$e^{-t\varepsilon c^*c} = -\varepsilon \int da^* da \ \exp[\ \frac{1}{\varepsilon} \ a^*a - \sqrt{t}(a^*c+c^*a)] \quad (2)$$

where a and a* are auxiliary fermion fields. Using (2) for
every Fourier component enables us to write the partition
function as a functional integral over both a- and c-fields.
In this theory, hopping is thus generated via a coupling to
the auxiliary fields.

Since the energy ε is scaled out in (2), the a-c
coupling term becomes site-diagonal. This allows us to in-
tegrate out the c-fields by expanding in powers of t, which
requires solving a one-site problem exactly for t = 0. The
result is a functional integral over the a-fields alone:

$$Z = Z_0 \pi(-\varepsilon_k) \int Da^* Da e^{S(a^*,a)} \quad (3)$$

where Z_0 is the partition function in the absence of hopping, and the new 'action' S is given by

$$S = [(t\varepsilon_k)^{-1} - g_0(\omega)]a_\sigma^*(k\omega)a_\sigma(k\omega)$$

$$+ \frac{t^2}{4N} \int_0^\beta \pi d\tau \Gamma_{\sigma\sigma'} a_\sigma^*(1)a_\sigma^*(2)a_\sigma^*(3)a_\sigma(4)+0(t^3) \quad (4)$$

In (4), $g_0(\omega)$ is the single-site Green's function for $t = 0$, Γ is the corresponding two-point function, $a(1)$ stands for $a(k_1, \tau_1)$, etc. Equation (4) can be considered as a series in powers of the number of hops which describe correlated multi-particle motion. The strong correlations appear through g_0 and Γ which contain the effect of U exactly.

The main quantity of interest is the exact single-particle Green's function $G_c(k\omega) = \langle c^*(k\omega)c(k\omega)\rangle$. This quantity can be obtained from the corresponding a-particle Green's function through the exact relation $G_c(k\omega) = G_a(k\omega) - 1/t\varepsilon_k$.

The first-order results are obtained by neglecting the t^2 term in (4). This gives $G_c^{-1}(k\omega) = g_0^{-1}(\omega) - t\varepsilon_k$ where

$$g_0 = (1-n_0/2)(i\omega+\mu)^{-1} + (n_0/2)(i\omega+\mu-U)^{-1} \qquad (5)$$

Here μ is the chemical potential, $n_0(\mu)$ is the density for $t = 0$. The first peak in g corresponds to having no particle, while the second peak corresponds to having another particle of opposite spin at a given site. For non-zero t, these two peaks broaden into two bands centered at $-\mu$ and $U-\mu$, respectively. For $n = 1$, these bands are separated by a correlation gap $\Delta_c = (D^2+U^2)^{1/2}-D$ where D is the bandwidth of the non-interacting system ($t = 0$). The first-order Green's function is thus the same as the one obtained by Hubbard.[5] It is gratifying to see that this physically reasonable two-band structure appears naturally as the lowest-order approximation to a systematic expansion in powers of t. Note also that for large U and $n = 1$, the ground state energy in this approximation is $E \simeq -$ const. D^2/U, and is thus already better than the Gutzwiller variational result.[6]

Large U and n = 1

It is possible to improve the first-order results by doing a Hartree-Fock decomposition of the–second order term in (4). However, for large U and n close to unity, the doubly occupied or the empty state at a given site is very rare. It is then possible to project out these states perturbatively in a coupled field theory by writing the coupling term as:

$$t\int d\tau d\tau' \phi(\tau-\tau') \ a_\sigma^*(\tau) f_\sigma(\tau) f_{\sigma'}^*(\tau') a_{\sigma'}(\tau') \tag{6}$$

where $\phi(\tau) = \theta(\tau)e^{-(U-\mu)\tau} + \theta(-\tau)e^{\mu\tau}$ is just the sum of the propagators for the doubly occupied and the empty states, respectively. In (6), the f-fields are constrained so that there is exactly one particle per site for n = 1. They are related to the original c-fields:

$$f_\sigma(\tau) f_{\sigma'}^*(\tau') \phi(\tau-\tau') \rightarrow c_\sigma(\tau) c_{\sigma'}^*(\tau')$$

The coupled functional integral involving the a- and f-fields is similar to the original coupled functional integral except for the constraints, provided that the hopping term is replaced by (6) and the U term of Eq. (1) is removed.

The interaction as represented by (6) corresponds to dynamic super-exchange between the a- and f- particles. We note that this coupled theory is equivalent to summing the perturbation series to all orders in t. If we now integrate out the auxiliary a-fields and let $\exp(-\alpha|\tau|) \rightarrow \delta(\tau)/\alpha$, we recover the usual super-exchange involving the f-fields alone. According to Anderson, it is this super-exchange which leads to pairing via the RVB mechanism.[2]

The effect of the RVB mechanism is to introduce new bands near $i\omega \sim 0$, band width being of order $(t\Delta)^2/U$, where the order parameter $\Delta \sim <f_\uparrow^* f_\downarrow^*>$ (see Ref. 2). However, in the usual super-exchange model, the two peaks obtained from the first-order approximation are lost, due to the delta-function approximation of ϕ. On the other hand, it is easy to see, by going to frequency space, that the two peaks are contained in our interaction term (6).

Our principal result is that the new bands near $i\omega = 0$ are in addition to the Hubbard bands near $i\omega = -\mu$ and $i\omega = U-\mu$, and that this new structure has a spectral weight which is of order $\Delta^2 \ll 1$. This is because most of the spectral weight (of order $-(1-\Delta^2)$) is carried by the two bands centered at $\pm U/2$ (since for $n = 1$, $\mu = U/2$) and is thus frozen out.

Useful conversation with P. B. Visscher is acknowledged. This research was supported by a summer grant from the University of Alabama Research Grant Committee.

REFERENCES

1. J. G. Bednorz and K. A. Müller, Z. Phys. B64, 188 (1986).
2. P. W. Anderson, Science 235, 1196 (1987); P. W. Anderson, G. Baskaran, Z. Zou, and T. Hsu, Phys. Rev. Lett. 58, 2790 (1987).
3. V. J. Emery, Phys. Rev. Lett. 58, 2794 (1987).
4. V. N. Popov, "Functional Integrals in Quantum Field Theory and Statistical Physics," D. Reidel Publishing Co., Dordrecht, (1983).
5. J. Hubbard, Proc. Roy. Soc. (London) A276, 238 (1963).
6. D. Vollhardt, Rev. Mod. Phys. 56, 99 (1984).

CHEMICAL BONDING TOPOLOGY OF SUPERCONDUCTORS

R. BRUCE KING
Department of Chemistry, University of Georgia, Athens,
Georgia, 30602, U.S.A.

Abstract   The high critical temperature copper oxide
superconductors exhibit porously delocalized chemical
bonding topologies closely related to those of the
ternary molybdenum chalcogenides (Chevrel phases)
and ternary lanthanide rhodium borides. The rigidities
of the conducting skeletons appear to play an important
role in determining critical temperatures.

INTRODUCTION

One of the most important scientific developments during
the past two years has been the discovery of copper oxide
superconductors having critical temperatures (T_c) much
higher than those of any previously known superconductors.[1]
Numerous theoretical approaches have been used to account
for these unprecedented superconducting properties.   This
paper summarizes a topological approach based on previous
studies[2,3,4,5,6] on metal cluster chemical bonding
topology.   The model is first developed using some of
the metal cluster superconductors with the highest T_c's
for structures of this type, namely the ternary molybdenum
chalcogenides (Chevrel phases)[5,7] and ternary lanthanide
rhodium borides.[5,8] Key ideas resulting from the study
of such metal cluster superconductors, such as the
occurrence of a porously delocalized conducting skeleton
in such systems, can then be applied to an analysis of
the high T_c copper oxide systems.[9]

359

METAL CLUSTER SUPERCONDUCTORS

The most important type of Chevrel phases have the general formulas $M_nMo_6S_8$ and $M_nMo_6Se_8$ (M = Ba, Sn, Pb, Ag, lanthanides, Fe, Co, Ni, etc.).[10,11] The basic building blocks of their structures consist of a bonded Mo_6 octahedron within an S_8 cube. The other metal atoms M furnish electrons to the resulting Mo_6S_8 units allowing them to approach, but not attain, the $Mo_6S_8^{4-}$ closed shell electronic configuration corresponding to a partially filled valence band. Electronic communication between individual Mo_6 octahedra is provided by interoctahedral metal-metal interactions. Closely related structures, also based on an Mo_6 octahedron within an X_8 cube, are found in the discrete molybdenum (II) halides $Mo_6X_8L_6^{4+}$, which have the closed–shell electronic configuration and no electronic communication between Mo_6 octahedra.[3,5]

The $Mo_6S_8^{4-}$ closed-shell configuration for the Chevrel phase building block has 24 skeletal electrons, which are the exact number required for an edge-localized octahedron having Mo-Mo two-center bonds along each of the 12 edges.[5,7] The chemical bonding topology of the Chevrel phases thus consists of edge-localized Mo_6 octahedra linked electronically into an infinite structure both through sulfur atoms and through interpolyhedral Mo-Mo interactions. This leads naturally to the concept of porous delocalization. Thus, the bonding in an edge-localized polyhedron is porous, whereas the bonding in a globally delocalized polyhedron is dense.[6] In topological terms porous chemical bonding uses a one-dimensional chemical bonding manifold corresponding to the 1-skeleton of the polyhedral network, in contrast to dense chemical bonding, which uses the whole volume of the polyhedral network. Relatively high superconducting T_c's for a given compound type can be associated with a porously delocalized conducting skeleton. This idea appears to be related to the suggestion[12] that the high critical field of the Chevrel phases arises from a certain localization of the conduction electron wave function in the Mo_6 clusters, leading to an extremely short mean free path and/or a low Fermi velocity, corresponding to a small B.C.S. coherence length.

These ideas are also supported by a study of the chemical bonding topology of the ternary lanthanide rhodium borides,[5,8] $LnRh_4B_4$ (Ln=lanthanides heavier than Nd), which exhibit significantly higher T_c's than other types of metal borides. These rhodium borides have a structure consisting of electronically linked Rh_4 tetrahedra similar to the discrete Rh_4 tetrahedron in the molecular species $Rh_4(CO)_{12}$.[5,8] Discrete B_2 units are arranged in this network of Rh_4 tetrahedra so that each triangular face of each Rh_4 tetrahedron is capped by a boron atom. The skeletal electron counting[5,8] corresponds to a closed-shell $Rh_4B_4^{4-}$ unit having Rh-Rh two-center bonds along each of the 6 edges of the Rh_4 tetrahedron as well as 12 Rh-B two-center bonds in each Rh_4B_4 unit. The lanthanide rhodium borides are thus porously delocalized systems like the Chevrel phases discussed above.

COPPER OXIDE SUPERCONDUCTORS

The ideas outlined above for metal cluster superconductors can be extended to high-T_c copper oxide superconductors by considering the following points:[9]

(1) The infinite conducting skeleton is constructed from Cu-O-Cu bonds rather than direct Cu-Cu bonds;

(2) The relevant metal-metal interactions are antiferromagnetic interactions between the single unpaired electrons of two d^9 Cu(II) atoms separated by an oxygen bridge similar to antiferromagnetic Cu(II)-Cu(II) interactions in discrete binuclear complexes[13], as well as resonating valence bond models;[14]

(3) The positive counterions in the copper oxide superconductors control the negative charge on the Cu-O skeleton, and thus the oxidation states of the copper atoms;

(4) Partial oxidation of some of the Cu(II) to Cu(III) generates the holes in the valence band required for conductivity.

These considerations lead to a porously delocalized chemical bonding topology for the high-T_c copper oxide superconductors, similar to the chemical bonding topologies of the Chevrel phases and lanthanide rhodium borides discussed above. The much higher ionic character and thus much lower polarizability and higher rigidity of metal-oxygen bonds (relative to metal-metal bonds) can then be related to the persistence of superconductivity in copper oxides to much higher temperatures than in metal clusters.

The rigidity of the Cu-O conducting skeleton also appears to be related to T_c in copper oxide superconductors. In the 40 K superconductors $La_{2-x}M_xCuO_{4-y}$ (M = Sr, Ba) the conducting skeleton consists of infinite single two-dimensional Cu-O layers of the stoichiometry CuO_{4-y}^{-6+x} separated by the positive counterions La^{3+} and

M^{2+}.[1,15] In the higher-T_c 90 K superconductors $YBa_2Cu_3O_7$ two similar two-dimensional Cu-O conducting skeletons are braced by perpendicular Cu-O chains leading to the stoichiometry $Ba_2Cu_3O_7^{3-}$ for the superconducting slabs, which are separated by the Y^{3+} counterions.[1,16] Thus although both the 40 K and 90 K superconductors have two-dimensional conducting skeletons, the braced conducting skeleton of the 90 K superconductor is more rigid than the unbraced conducting skeleton of the 40 K superconductor. This supports the role of the rigidity of the conducting skeleton in determining T_c in superconductors. Thus M-M skeletons (e.g., metal clusters such as the Chevrel phases and ternary lanthanide rhodium borides) have the lowest T_c's among the compounds considered in this paper, simple two-dimensional Cu-O-Cu skeletons (e.g., $La_{2-x}M_xCuO_{4-y}$) have intermediate T_c's, and braced Cu-O-Cu two-dimensional skeletons (e.g., $YBa_2Cu_3O_7$) have the highest T_c's. The positive counterions are therefore seen to have the role of determining the rigidity of the Cu-O conducting skeleton as well as the copper oxidation states.

ACKNOWLEDGEMENT

I am indebted to the U.S. Office of Naval Research for partial support of this research.

REFERENCES

(1)   J.M. Williams, M.A. Beno, K.D. Carlson, U. Geiser, H.C. Ivy Kao, A.M. Kini, L.C. Porter, A.J. Schultz, R.J. Thorn, H.H. Wang, M.-H. Whangbo, and M. Evain, Accts. Chem. Res., 21, 1 (1988).
(2)   R.B. King and D.H. Rouvray, J. Am. Chem. Soc., 99, 7834 (1977).

(3)   R.B. King in Chemical Applications of Topology
       and Graph Theory (R.B. King, Ed.), Elsevier,
       Amsterdam, 1983, pp. 99-123.
(4)   R.B. King in Molecular Structure and Energetics
       (J.F. Liebman and A. Greenberg, Eds.), Verlag Chemie,
       Deerfield Beach, Florida, 1986, pp. 123-148.
(5)   R.B. King in Graph Theory and Topology in Chemistry
       (R.B. King and D.H. Rouvray, Eds.), Elsevier,
       Amsterdam, 1987, pp. 325-343.
(6)   R.B. King, J. Math. Chem., 1, 249 (1987).
(7)   R.B. King, J. Solid State Chem., 71, 224 (1987).
(8)   R.B. King, J. Solid State Chem., 71, 233 (1987).
(9)   R.B. King, Inorg. Chim. Acta, 143, 15 (1988).
(10)   Ø. Fischer, Appl. Phys., 16, 1 (1978).
(11)   R. Chevrel, P. Gougeon, M. Potel, and M. Sergent,
       J. Solid State Chem., 57, 25 (1985).
(12)   Ø. Fischer, M. Decroux, R. Chevrel, and M. Sergent
       in Superconductivity in d- and f-Band Metals (D.H.
       Douglass, Ed.), Plenum Press, New York, 1976, pp.
       176-177.
(13)   C.J. Cairns and D.H. Busch, Coord. Chem. Rev.,
       69, 1 (1986).
(14)   P.W. Anderson, Science, 235, 1196 (1987).
(15)   M.-H. Whangbo, M. Evain, M.A. Beno, and J.M.
       Williams, Inorg. Chem., 26, 1829 (1987).
(16)   M.-H. Whangbo, M. Evain, M.A. Beno, and J.M.
       Williams, Inorg. Chem., 26, 1831 (1987).

CHARGE CIRCULATION IN HIGH-T$_c$ SUPERCONDUCTORS

ROGER E. CLAPP
The MITRE Corporation, Bedford, Massachusetts, U.S.A.

Abstract   The short coherence lengths in the copper
oxide superconductors imply that the active charges are
localized in configuration space. Yet these active
charges have substantial kinetic momenta **k**. Accordingly,
a moving charge may be executing a quantized orbital
motion in two dimensions. In the CuO "planes" the orbits
can be square or rectangular with lattice recoil at each
corner. In the CuO "chains" (found in the 1-2-3
compounds) the orbits can be long rectangles utilizing
O–O bridges between adjacent chains. Crystallography
shows that each of the chain oxygens is in motion.
Photoemission spectra show that holes in the oxygen 2p
valence band give rise to resonating O–O bonds. If this
long rectangular orbit has Möbius topology, then the
stored kinetic energy matches $k_B T_c$ for $T_c \sim 90$ K when
the orbit is 38 bonds long, with its period synchronized
to an oxygen libration at the observed Raman frequency
of 483 cm^{-1}.

INTRODUCTION

From the measured structure of YBa$_2$Cu$_3$O$_{7-\delta}$ (Figure 1) it is
evident that most Cu–O bonds are too short to be strictly
ionic, and must be at least partially covalent. Oxygen
itself is bivalent, using p-electron σ-bonds that cannot be
collinear. In the CuO planes containing atoms designated as
Cu(2), O(2), and O(3), the oxygen bond angle is kept less
than 180° by puckering: Cu(2)——O(2,3)——Cu(2)

while in the CuO chains, containing Cu(1) and O(1), the O(1)
atoms have a large anisotropic ellipsoid of uncertainty,[1]
showing continuous motion: Cu(1)< ⩵O(1)⩳ >Cu(1)

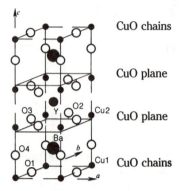

FIGURE 1   Crystal structure of $YBa_2Cu_3O_{7-\delta}$. Reprinted by permission from David et al., <u>Nature</u>, <u>327</u>, 310. Copyright (c) 1987 Macmillan Journals Limited.

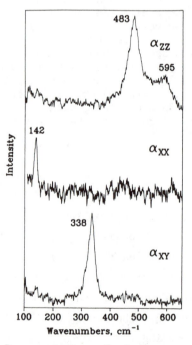

FIGURE 2   Raman spectra of single crystals of $YBa_2Cu_3O_{7-\delta}$. Reprinted by permission from Hemley and Mao, <u>Phys. Rev. Letters</u>, <u>58</u>, 2340 (1987).

CHARGE CIRCULATION IN HIGH-T$_c$ SUPERCONDUCTORS

VIBRATIONAL MODES

$$
\begin{array}{c}
O(4) \\
| \\
Cu(1) \\
| \\
O(4)
\end{array}
$$

In Figure 2 the top spectrum has the polarizations of the incident and scattered radiation both parallel to the c-axis, hence normal to the planes and chains.[2] The 483 cm^{-1} line is the Cu(1)-O(4) stretching vibration.[3]

The polarizations in the lower spectra in Figure 2 are in the a- and b-directions, perpendicular to the c-axis. In particular, the 338 cm^{-1} line represents a bending vibration of the Cu(2)-O(2) and Cu(2)-O(3) bonds in the CuO planes.[3]

OXYGEN-OXYGEN BONDING

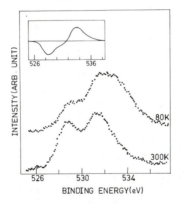

FIGURE 3   X-ray photoelectron spectra of O(1s) in YBa$_2$Cu$_3$O$_{7-\delta}$ at 300 K and 80 K. Reprinted by permission from Sarma et al., **Phys. Rev. B, 36**, 2371 (1987).

The inset shows the difference between the spectra recorded at 80 and 300 K. The growth of an oxygen dimer signal at 533.3 eV is seen to be accompanied by a reduction in the oxygen monomer signal at 528.7 eV. This is evidence of oxygen-oxygen coupling for T less than T$_c$ ~ 90 K.

367

LOOP CURRENTS

The oxygen–oxygen bonding in Figure 3 has been attributed to holes in the oxygen 2p valence band.[4] We can note that the only pairs of oxygens in the $YBa_2Cu_3O_{7-\delta}$ crystal (Figure 1) that can get close enough together to be momentarily bonded are the O(1) atoms in adjacent chains. We suggest that the O(1) atoms are in librational motion, during which they swing around in oval paths. Furthermore, we suggest that the O(1) libration is synchronized to the Cu(1)–O(4) vibration shown in the top curve in Figure 2. The motion is such that the Cu(1)–O(1)–Cu(1) bonding never becomes collinear, and the O(1) ellipsoid of uncertainty is thereby understood.

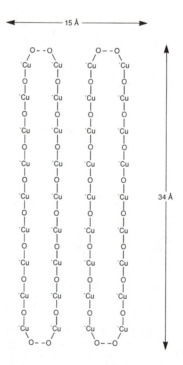

FIGURE 4   Two adjacent orbital loops formed from CuO chain segments linked by momentary O–O bridges.

CHARGE CIRCULATION IN HIGH-T$_c$ SUPERCONDUCTORS

Localization of an itinerant hole to the dimension of
the observed coherence length of 34 Ångstroms, normal to
the c–axis,[5] can be accomplished if the hole is executing a
circulation around one of the orbits shown in Figure 4.
An 0–0 bridge will form every 69 femtoseconds (fs) if the
two oxygen atoms are in synchronized libration and if a
circulating hole is synchronized to cross the bridge
repeatedly and provide the binding energy for stabilization.

A further synchronization can couple together two holes,
orbiting separately in the adjacent loops shown in Figure 4.
When a circulating positive charge reaches one end of a
loop, the momentum reversal results in a recoil momentum
absorbed by the crystal lattice. If there is a simultaneous
reflection of the second charge from the opposite end of the
adjacent loop, then the two oppositely-directed impulses, if
synchronized to a lattice vibration, will counteract,
minimizing the propagation of pulses of vibrational energy
away from the pair of loops. The phonon energy thus saved
can act to bind the two circulating charges into a Cooper
pair. The coupled synchronization of many Cooper pairs can
reduce still further the dissipation of vibrational energy
while maintaining the storage of kinetic energy in the
motion of the circulating electronic charges.

If the ambient thermal quanta are too weak to disturb
the loop currents, the synchronization of the coupled loop
excitations will be preserved. But when the thermal quanta
$k_B T$ exceed a critical value $k_B T_c$ there will be disturbances
that uncouple the Cooper pairs. The transition comes when

$$k_B T_c = p^2/2m, \tag{1}$$

where m is the effective mass, while p is the momentum which
is related to the de Broglie wavelength λ_{deB} through

$$p = h/\lambda_{deB}. \tag{2}$$

The de Broglie wavelength, in turn, can be obtained from the perimeter of the orbital loop, and is in fact equal to twice that perimeter if the orbit contains a Möbius phase reversal[6-8] and is in its lowest state of excitation. This lowest state has the orbital angular momentum quantum number

$$L = 1/2, \tag{3}$$

and is metastable against a radiative loss of energy, since a radiative process ordinarily requires a transition with

$$\Delta L = 1. \tag{4}$$

MAGNETISM

The circulation with $L = 1/2$ will have an attached magnetic field of one-half flux line, which can aid in the coupling between adjacent loops. When magnetism is present, we will want to write

$$p = k - (e/\hbar c)A \tag{5}$$

where k is the kinetic momentum and A is the vector potential. The London equation

$$J = -(e^2 n/mc)A \tag{6}$$

relates a supercurrent J to a vector potential A. Substitution in Eq. (5) then gives

$$\delta k = (m/ne\hbar)J, \tag{7}$$

where n is the charge carrier density. The increment δk represents a displacement of the Fermi surface in momentum space, with a consequent net current flow.[9] The value of δk can be controlled by the chains while the supercurrent actually flows mainly in the planes.

CHARGE CIRCULATION IN HIGH-T$_c$ SUPERCONDUCTORS

MÖBIUS ORBITS

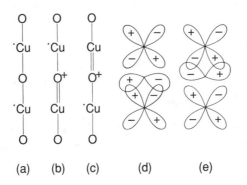

(a)	(b)	(c)	(d)	(e)

FIGURE 5   The phase reversal in a Möbius orbit.

In Figure 5, (a) is a portion of a CuO chain from Figure 4;
(b) shows the insertion of a positive charge converting a
bivalent O to a trivalent O^+, coupled by a π-bond to the
unpaired d-electron in an adjacent Cu; (c) shows the same O^+
coupled to the other adjacent Cu; (d) pictures the bonding
in (b), with the p-electron phased to lean toward the lower
d-electron; while (e) pictures the bonding in (c), with a
phase reversal that makes the p-electron lean upward.

The migrating p-electron hole travels down one chain,
crosses over an O–O bridge, travels up the adjacent chain,
crosses back on the upper O–O bridge, then travels down the
first chain and re-enters the original oxygen P-state but
with its phase reversed. For repeated orbital circulation we
require that the wave function have phase continuity. Here
this means that there needs to be an additional 180° of
phase shift provided by travel over a path length of one-
half de Broglie wavelength. That is, the perimeter of the
Möbius orbit equals $\lambda_{deB}/2$, so that we obtain λ_{deB} by
doubling the orbital perimeter.

371

In $YBa_2Cu_3O_{7-\delta}$ there is a plateau[10] for $0<\delta<0.2$, with $T_c \sim 90$ K. For this plateau we find that k_BT_c roughly equals the kinetic energy for a 38-bond Möbius orbit (Figure 4). The orbital travel period (142 fs) is approximately twice the postulated oxygen libration period (69 fs), providing synchronization between charge circulation and 0–0 bridgings. A small reduction in the period for 0–0 bridgings should have little effect on T_c, but reductions that increased the matching orbits to 42 or 46 bonds would lower T_c to 74 K or 61 K, respectively. Compare this to the result of replacing O^{16} by O^{17} and O^{18}. These substitutions have been reported[11] to lower T_c from 92 K to 77 K and 59 K.

In $YBa_2Cu_3O_{7-\delta}$ there is also a 60 K plateau[10] for $0.3<\delta<0.4$, where chain connectivity is lost.[12] All the currents must be flowing in the planes, where we can have a squared Möbius orbit that is 44 bonds in perimeter, giving $T_c \sim 60$ K. The orbital period is 200 fs, about twice the 99 fs period of the 338 cm^{-1} vibration shown in Figure 2.

REFERENCES

1. W. I. F. David et al., Nature, 327, 310 (1987).
2. R. J. Hemley and H. K. Mao, Phys. Rev. Letters, 58, 2340 (1987).
3. H. J. Rosen et al., Bull. Am. Phys. Soc., 33, 380 (1988).
4. D. D. Sarma et al., Phys. Rev. B, 36, 2371 (1987).
5. T. K. Worthington, W. J. Gallagher, and T. R. Dinger, Phys. Rev. Letters, 59, 1160 (1987).
6. E. Heilbronner, Tetrahedron Letters, 29, 1923 (1964).
7. H. E. Zimmerman, Quantum Mechanics for Organic Chemists (Academic Press, New York, 1975).
8. R. E. Clapp, Theoret. Chim. Acta, 61, 105 (1982).
9. J. M. Ziman, Principles of the Theory of Solids, second edition (Cambridge University Press, 1972), pp. 394-397.
10. R. J. Cava et al., Nature, 329, 423 (1987).
11. K. C. Ott et al., submitted.
12. W. W. Warren, Jr., Bull. Am. Phys. Soc., 33, 733 (1988).

THEORETICAL STUDIES OF CuO CLUSTERS REPRESENTING YBa$_2$Cu$_3$O$_{7-x}$: INVESTIGATION OF Cu(III) FORMATION VERSUS OXIDE OXIDATION*

L. A. CURTISS and A. SHASTRI[†]
Chemical Technology Division/Materials Science Program
Argonne National Laboratory
Argonne, Illinois 60439-4837

Abstract *Ab initio* molecular orbital calculations on a simple CuO system, $(CuOH_2)^{3+}$, indicate that oxygen hole (pπ) formation is favored over Cu(III) formation. Relation of these results to YBa$_2$Cu$_3$O$_{7-x}$ is discussed.

INTRODUCTION

The discovery of the copper oxide superconductors with high critical temperatures, T$_c$, has caused a great deal of speculation as to the mechanism of superconductivity in these new materials. From the beginning, it was recognized that the stoichiometry required a mixture of formal oxidation states of copper or oxygen.[1-6] The copper oxides layers in the undoped systems, YBa$_2$Cu$_3$O$_{6.5}$ and La$_2$CuO$_4$, formally contain Cu's and O's with 2+ and 2− charges respectively. When the material is doped either by M = Ba(Sr) in La$_{2-x}$M$_x$CuO$_4$ or by adding oxygen in YBa$_2$Cu$_3$O$_{7-x}$ electrons are removed from

*Work supported by the U.S. Department of Energy, Division of Materials Sciences, Office of Basic Energy Sciences, under Contract W-31-109-ENG-38.

[†]Graduate Student Participant from Physics Department, Carnegie-Mellon University, Pittsburgh, Pennsylvania.

the copper or oxygen giving rise to mixed oxidation states. There have been conflicting reports, both experimental and theoretical, as to whether electron removal from the copper (i.e Cu(III) formation) or the oxygen (i.e. oxide formation) occurs.

We have carried out *ab initio* molecular orbital calculations for some small CuO systems, $[Cu(OH_2)_n]^{m+}$, to determine if the electron comes from an oxygen or copper orbital when the charge is increased from 2+ to 3+. This is a simple representation of the "doping" process in the CuO chains and planes in $YBa_2Cu_3O_{7-x}$. In this paper we report on the results for n = 1 with a summary of some of the key results for n = 4 included in the discussion section.

THEORETICAL METHODS

The *ab initio* molecular orbital theory used is single determinant SCF-LCAO-HF-MO theory[7-9] with all electrons included. We have used $[Cu(OH_2)_n]^{m+}$ clusters to represent the Cu$-$O interaction in the copper oxide layers with the hydrogens truncating the oxygen. Other methods of truncation including use of Be or point charges give results similar those presented here using hydrogen truncation. There is reason to believe that $[Cu(OH_2)_n]^{m+}$ clusters can represent to some extent the local electronic structure of the copper oxide in these materials as x-ray diffraction studies[10] of Cu^{2+} in liquid water give a Jahn-Teller distortion similar to that found in $La_{2-x}M_xCuO_4$. Also note that our theoretical method includes electrostatic, covalent, polarization, etc. forces at the first principles level.

RESULTS

Calculations on $(CuOH_2)^{2+}$ indicate that it is characterized by an interaction of a Cu^{2+} d^9 ion and the oxygen of H_2O with some charge transfer from the oxygen lone pair to the Cu. Removal of an electron from $(CuOH_2)^{2+}$ can come from either the copper or oxygen orbitals to give $(CuOH_2)^{3+}$. Two potential energy curves for the $(CuOH_2)^{3+}$ cluster[11] are illustrated in Fig. 1. The geometrical configuration corresponds to a C_{2v} planar structure with the hydrogens pointing away from the Cu (the OH distance is 0.958 Å and the HOH angle is 104.5°). Of the two potential energy curves the higher (A) has an attractive well and corresponds to interaction of Cu^{3+} with OH_2. This represents Cu(III) and unoxidized oxygen, i.e., Cu^{3+} + O^{2-}. The lower curve (B) is repulsive and corresponds to interaction of Cu^{2+} with $^+OH_2$. This represents Cu(II) plus an oxygen hole (O_{hole}), i.e, Cu^{2+} + O^-. At the distances (1.9 Å) found in the copper oxide planes and chains the latter interaction is favored by about 8 eV. The large energy difference is due in part to the large third ionization potential of copper which makes removal of the third electron from Cu difficult energetically.

The wave function for curve B indicates that the electron comes out of a localized $p\pi$-type oxygen orbital. A configuration having the electron removed out of a σ-type oxygen orbital is 3.9 eV higher in energy. If one takes into account the fact that the energy difference between the π (b_1) and σ (a_1) orbitals of H_2O is 1.9 eV at this level of calculation, then the actual energy difference between the π and σ hole states due to interaction with Cu is 2.0 eV. Also, the O $2p\pi$-hole state is a triplet, so that the single oxygen π-type electron is spin–unpaired with the single Cu electron in a σ-type orbital.

375

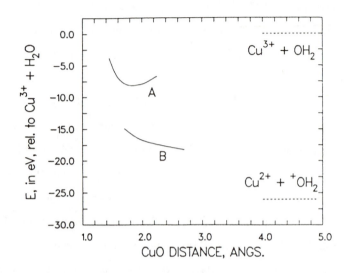

FIGURE 1 Potential energy curves for the interaction of Cu^{3+} and OH_2 (curve A) and the interaction of Cu^{2+} with $^+OH_2$ (curve B). On the right hand side are shown energies of separated species.

DISCUSSION AND CONCLUSIONS

These results can be used to help understand the Cu valency in copper oxides. When more H_2O's are added to $(CuOH_2)^{3+}$ to form $[Cu(OH_2)_n]^{3+}$, the energy difference between the Cu(III) state (curve A) and the oxygen hole state (curve B) will become smaller because of a deeper attractive well in the former case. (An attractive potential well also occurs for the oxygen hole state if the hole is delocalized over the H_2O's.) In terms of copper oxide lattices this means that higher oxygen coordination of Cu and shorter CuO distances stabilizes Cu(III) relative to Cu(II) + O_{hole}. These conclusions are consistent with valence concepts in inorganic chemistry.[12]

$YBa_2Cu_3O_{7-x}$ has unusually short CuO distances in the CuO_3

chains (the $Cu(1)-O(4)$ distance is 1.84 Å, while the $Cu(1)-O(1)$ distance is 1.94 Å); but it is not quite equivalent to $KCuO_2$ which is often claimed to contain trivalent copper[13] (all $Cu-O$ distances are 1.84 Å). The question of whether $Cu(III)$ will be stable relative to $Cu(II) + O_{hole}$ when there is oxygen coordination as in $YBa_2Cu_3O_{7-x}$ cannot be answered with absolute certainty by our cluster calculations; however, in calculations on Cu coordinated by four waters, i.e., $[Cu(OH_2)_4]^{3+}$, with CuO distances similar to that in the superconducting materials, we find $Cu(II)$ with an oxygen hole in the $p\pi$-type orbital to be the stable state. This is in agreement with the cluster studies of Goddard et. al.[5] who also have found that the doped copper oxides contain $Cu(II)$ plus oxygen in-plane $p\pi$ holes as opposed to mixed valent $Cu(II)/Cu(III)$. We have also investigated $[Cu(OH_2)_4]^{2+}$ which represents undoped copper oxide and find formation of $Cu(I) + O_{hole}$ to be significantly higher in energy than $Cu(II)$ and unoxidized oxygen.

In conclusion, *ab initio* molecular orbital calculations for a simple CuO system $(CuOH_2)^{3+}$ indicate that oxygen hole $(p\pi)$ formation is favored over $Cu(III)$ formation by about 8 eV due to the large third ionization potential of Cu atom. Increasing the coordination of Cu by H_2O's decreases the energy difference. The results suggest that the "doped" copper oxide chains and planes in $YBa_2Cu_3O_{7-x}$ may be characterized by oxygen $(p\pi)$ holes and $Cu(II)$ rather than $Cu(III)$. As a consequence, mixed valency in $YBa_2Cu_3O_{7-x}$ as well as other CuO containing materials could give rise to oxygen valence fluctuations $(O^{2-} \rightleftarrows O^-)$ which may be important in the mechanism of superconductivity. Future work will consider larger clusters, other transition metals, and implications of these oxygens holes in terms of the mechanism of superconductivity.

REFERENCES

1. M. K. Wu, J. R. Ashburn, C. J. Torng, P. H. Hor, R. L. Meng, L. Gao, Z. Huang, Y. Q. Wang, C. W. Chu, Phys. Rev. Lett. 58, 908 (1987).

2. V. J. Emery, Phys. Rev. Lett. 58, 2794 (1987).

3. D. C. Mattis and M. P. Mattis, Phys. Rev. Lett. 59, 2780 (1987).

4. F. J. Adrian, Phys. Rev. B37, 2326 (1988).

5. a) Y. Guo, J. M. Langlois, and W. A. Goddard, Science 239, 896 (1988); b) G. Chen and W. A. Goddard, ibid , p. 899.

6. L. A. Curtiss, T. O. Brun, D. M. Gruen, Inorg. Chem. 27, 1421 (1988).

7. W. J. Hehre, L. Radom, J. A. Pople, P. v. R. Schleyer, "Ab Initio Molecular Orbital Theory," John Wiley (New York, 1987).

8. The Cu, O, H basis sets are 9s3p2d, 3s2p, and 2s contractions, respectively, of primitive gaussians similar to that used for Fe, O, and H in Ref. 9.

9. L. A. Curtiss, J. W. Halley, J. Hautman, and A. Rahman, J. Chem. Phys. 86, 2319 (1987).

10. Y. Marcus, "Ion Solvation," John Wiley (New York, 1985).

11. In fig. 1 the lower curve (B) corresponds to a $^3B_1$ state with a total energy of -1713.015502 a.u. at 1.9 Å. The upper curve (A) was estimated on the basis of calculated potential energy curves for $Fe^{3+}-OH_2$ (ref. 9), $Fe^{2+}-OH_2$ (ref. 9), and $Cu^{2+}-OH_2$ (to be published).

12. F. A. Cotton and G. Wilkinson, "Advanced Inorganic Chemistry," (Interscience, New York, 1966).

13. K. Hestermann and R. Hoppe, Z. Anorg. Allg. Chem. 367, 249 (1969).

A MOLECULAR ORBITAL STUDY OF BONDING IN $YBa_2Cu_3O_7$

E.A. Boudreaux and L. Kwark
Department of Chemistry University of New Orleans New Orleans, Louisiana 70148, U.S.A.

Abstract Preliminary M.O. calculations have been carried out on CuO clusters in one vertical stack and one horizontal sheet of the $YBa_2Cu_3O_7$ structure. The results show that oxidation from oxygen, and not copper, is most likely to take place.

INTRODUCTION

Preliminary molecular orbital calculations utilizing the S.C.M.E.H. - M.O. method[1-5] have been carried out on CuO cluster components of the high T_c superconductor $YBa_2Cu_3O_7$. One Cu_4O_{14} cluster in the vertical stacks ("V" cluster) and one Cu_4O_{16} cluster in the horizontal sheets ("H" cluster) have been treated as separate molecular units, and the results compared.

The internuclear distance adopted for these calculations are: "V" cluster - $R(CuO)_{eq} = 1.930Å$; $R(CuO)_{ax} = 2.295Å$. Assuming the "V" cluster to contain Cu(III) yields a cluster charge of -16, while a charge of -24 must be assigned to the "H" cluster containing Cu(II). Both of these charges are much too high to allow convergence of the calculations. This is in spite of the fact that the computational format provides a counter - charge compensating potential. Actually the S.C.M.E.H. - M.O. method does not require any specification of charged states. Only the total electron count on each atom, associated orbital energy and overlap parameters, and the net cluster charge (if any) need be specified. The calculation is thus carried to self-consistency in both net atomic charges and orbital configurations.

In order to counteract the non-convergence problem due to high charges, and still preserve the cluster ionicity, Mg counter atoms were placed at the standard ionic Mg-O bond distances of 2.104Å from the non-bridging oxygens. Stoichiometric formulations $Mg_8Cu_4O_{14}$ and $Mg_{12}Cu_4O_{16}$ were assumed, and the respective orbitals and electrons: $Cu(3d4s4p)^{11}$, $O(2s2p)^6$ and $Mg(3s)^2$ employed, thus giving total electron counts of 144 and 164 for "V" and "H" clusters respectively. Other pertinent input parameters were derived from HF-SCF atomic calculations, as discussed elsewhere.[1,2]

RESULTS AND DISCUSSION

Calculated net atomic charges and orbital populations are presented in Table I. It is notable that the net charge on Cu in the "V" cluster (+0.55) is actually somewhat smaller than that for the "H" cluster (+0.60). Both of these are substantially smaller than Cu charges attained in other calculations.[6-9] Thus Cu(III) in the "V" cluster is not confirmed by these results. Using standard relations between electronegativity, ionic character and charge distribution,[10] an estimated Cu charge of +0.6 is obtained for the Cu-O bond. A charge much higher than this is an unrealistic reflection of excessive bond polarity.[11] Also, the charge in the "V" cluster axial O atoms is some 3.6 times more negative than that of the "H" cluster O atoms. Since these axial O atoms are the same when both clusters are combined, there is apparently a net transfer of electronic charge from the "H" to the "V" cluster. Thus, it can be deduced from these calculated net cluster charges, that effectively 0.188 units of electronic charge are transferred between the "H" and "V" clusters.

The M.O. eigenvalues and percent A.O. character per M.O. are presented in Tables II and III. These data show that the majority of M.O.'s are energetically very closely spaced. In the "H" cluster M.O. 43 is the HOMO with 1.804 electrons, and composed entirely of O 2_{px}, 2_{py} character. The LUMO is only some 0.05 eV higher, and is primarily O 2_{py} character. The nearest vacant M.O. with Cu character

M.O. STUDY OF BONDING IN $YBa_2Cu_3O_7$

TABLE I

ATOMIC CHARGES AND POPULATIONS "V" CLUSTER:

"V CLUSTER":

Atoms	Net Atomic Charge[*]		Orbital Populations					
	$Mg_8Cu_4O_{14}$	$Cu_4O_{14}^{-0.008}$	3d	4s	4p	2s	2p	3s
Cu	+0.554	+0.554	9.73	0.36	0.36	--	--	--
$O_1O_2^a$	-0.597	-0.597	--	--	--	1.85	4.60	--
O_3-O_6^b	-0.630	-0.165	--	--	--	1.90	4.21	--
O_7-O_{10}^c	-0.512	-0.047	--	--	--	1.90	4.04	--
O_{11}-O_{14}^c	-0.510	-0.045	--	--	--	1.90	4.04	--
Mg_{18}-Mg_8	+0.698	--	--	--	--	, --	--	1.30

"H" CLUSTER:

Atoms	Net Atomic Charge[*]		Orbital Populations					
	$Mg_{12}Cu_4O_{16}$	$Cu_4O_{16}^{+0.196}$	3d	4s	4p	2s	2p	3s
Cu	+0.604	+0.604	9.76	0.32	0.32	--	--	--
O_1-O_4^a	-0.388	-0.388	--	--	--	1.85	4.39	--
O_5-O_8^b	-0.481	-0.046	--	--	--	1.90	4.03	--
O_9-O_{12}^c	-0.513	-0.081	--	--	--	1.90	4.09	--
O_{13}-O_{16}^c	-0.518	-0.086	--	--	--	1.90	4.10	--
Mg_1-Mg_{12}	+0.432	--	--	--	--	--	--	1.57

*Counter Mg atoms removed and all electron density due to Mg is subtracted from O atoms in Cu_4O_{14} and $Cu_4 O_{16}$ clusters; a) bridging Cu-O-Cu; b) axial; c) terminal.

is 8.1 eV higher, and totally $4p_z$. The nearest occupied Cu containing M.O. is 71, 2.4eV below the HOMO. It is 45% Cu $3d_{x^2-y^2}$ and 55% O $2s$ $2p_x$, $2p_y$ character.

By contrast the "V" cluster is rather different. The HOMO is still $O2p_x$ character, and the LUMO is nearly degenerate with the HOMO, having $O2p_y$, $2p_z$ character. The nearest unoccupied Cu containing M.O. is 4.9 eV higher, and totally $4p_x$ character. The

381

nearest occupied M.O. with non-negligible Cu character, is some 4.1 eV below the HOMO.

A correlation of these results with the observed magnetic susceptibility data has also been made. The reported susceptibility[12] of 5×10^{-4} emu per mole for $YBa_2Cu_3O_7$ at 90K, yields a magnetic moment, $\mu_{eff} = 0.6$ B.M.. If the "spin-only" formula is applied, an effective spin of 0.09 is obtained. The 0.188 units of electrons transferred between "H" and "V" clusters provides an effective spin of 0.094.

TABLE II
EIGENVALUES AND PER CENT A.O. CHARACTER "H" CLUSTER

M.O.	Occupancy	Eigenvalues[*] (eV.)	Per Cent A.O. Character
16	0	2.686	Cu(4pz)~100%
17	0	-3.414	O(2pz) > 94%
↓			
42(LUMO)	0	-5.419	O(2py) 83%; O(2pz) 17%
43(HOMO)	1.804	-5.470	O(2py) 40%; O(2px) 45%
↓	˙(2)		O(mixed 2p's) 15%
70	2	-7.825	O(2pz)~ 100%
71	2	-7.907	Cu(3d x²-y²) 45%; O(2s, 2p) 55%
72	2	-7.918	O(2pz) 96.4%; Cu(3dz²) 2.7%; Cu(4pz) 0.9%.
73	2	-8.764	O(2s, 2px) 79%; Cu(3dx²-y²) 19% Cu(3dz²) 2.0%
74	2	-8.834	O(2px,y) 79.5%; Cu(3dz²) 18.8%
↓	↓		Cu(4s) 1.7%
79	2	-9.494	Cu(3dz², 3dx²-y²)~24-68%
↓	↓		
95	2	-9.705	Cu(3d's) 770-100%
↓	↓		
97	2	-34.751	O(2p's) ~100%

[*] Not corrected for spin-pairing and ligand field effects.

M.O. STUDY OF BONDING IN $YBa_2Cu_3O_7$

TABLE III

EIGENVALUES AND PER CENT A.O. CHARACTER "V" CLUSTER

M.O.	Occupancy	Eigenvalue* (eV.)	Per Cent A.O. Character
16	0	-1.170	Cu(4px) ~ 100%
↓			
35(LUMO)	0	-6.088	O(2py,z) >83%; (mixed 2p's)17%
36(HOMO)	0.008	-6.096	O(2px) >72%; (mixed 2p's) 28%
↓	(2)		
	↓		
62	2	-8.460	Cu(3dyz) 4.2%
63	2	-9.475	Cu($3dz^{2)}$) 3.4%; Cu($3dx^2$-y^2) 1.7%; Cu (4s) 1.6%
64	2	-9.475	Cu($3dz^2$, $3dx^2$-y^2) 12.5%
65	2	-10.266	Cu($3dz^2$, (3dyz) 15.6%
66	2	-10.304	Cu(3dyz) 18.2%
67	2	-10.541	Cu($3dz^2$, $3dx^2$-y^2) . 75-100%
↓	(2)		↓
	↓		
86	2	-11.451	
87	2	-34.490	O(2ps) ~ 100% p's

* Not corrected for spin-pairing and ligand field effects.

CONCLUSION

The results obtained in these preliminary calculations do not support the general view that mixed Cu^{2+}/Cu^{3+} oxidation states are involved in superconductivity. Rather, it appears that the oxidation of O is the required initial step of the mechanism, in agreement with Goddard.[13]

More refined calculations are currently in progress an a Cu_4O_{14} - Cu_4O_{16} coupled cluster. The results will be reported elsewhere.

E.A. BOUDREAUX AND L. KWARK

ACKNOWLEDGEMENTS:

The use of the UNO Computer Research Center and the Chemistry Department assistance is gratefully appreciated.

REFERENCES

1. E.A. Boudreaux, S.P. Doussa, and M. Klobukowski, Int. J. Quant. Chem.: Quant. Chem. Symp. 20, 239 (1986).
2. E.A. Boudreaux, Inorg. Chim. Acta. 82, 183 (1984).
3. T.P. Carsey and E.A. Boudreaux, Theoret. Chim. Acta. 56, 211 (1980).
4. L.E. Harris and E.A. Boudreaux, Inorg. Chim. Acta. 9, (1974)
5. A. Dutta-Ahmed and E.A. Boudreaux, Inorg. Chem. 12, 1590 (1973); Ibid, 12, 1597 (1973).
6. W.Y. Ching, Y. Xu, G.-L. Zhao, K.W. Wong and F. Zandichuadem, Phys. Rev. Lett. 59, 1333 (1987).
7. E.E. Alp, G.K. Shenoy, D.G. Hinks, D.W. Capone II, L. Soderholm and H.B. Schuttler, Phys. Rev. B35, 7199 (1987).
8. H. Chen, J. Calloway and P.K. Misra Phys. Rev. B36, 8863 (1987).
9. L.A. Curtiss. Private communication of Extended Hückel results prior to publication in Inorg. Chem.
10. C.A. Coulson, Valence, Oxford Press London, 1956, pp 102-105; 128-135.
11. R.T. Sanderson, Polar Covalence, Academic Press, New York, 1983, chaps 3, 15, 16. The calculated charge for Cu in the "H" cluster using Sanderson's approach is +0.50.
12. S. Davison, et al. in Chemistry of High Temperature Superconductors, (D.L. Nelson, M.S. Whittingham and T.F. George, eds.) ACS Symp. Series 351, Washington, D.C., 1987, p. 76.
13. W.A. Goddard III, Science, 239, 896(1988).

AUTHOR INDEX

AFFILIATION INDEX

Pennsylvania State University, University Park, Pennsylvania,165
Princeton University, Princeton, New Jersey, 69

San Jose State University, San Jose, California, 125

Texas Tech University, Lubbock ,Texas, 125

University of Alabama in Huntsville, Huntsville, Alabama,   23
University of Alabama, Tuscaloosa, Alabama, 151, 159, 353
University of Arkansas, Fayetteville, Arkansas, 175
University of Cincinnati, Cincinnati, Ohio, 125, 173
University of Electro-Communications, Tokyo, Japan, 119
University of Georgia, Athens, Georgia, 301, 359
University of Hawaii, Honolulu, Hawaii, 113
University of Houston, Houston, Texas, 173
University of New Orleans, New Orleans, Louisiana, 379
University of North Carolina, Chapel Hill, North Carolina, 235
University of Tokyo, Tokyo, Japan, 37
Ukrainian Academy of Sciences, Kharkov, USSR, 349